SCI PUBLICATION P334

Design of multi-storey braced frames

D G Brown BEng CEng MICE

C M King BSc MSc DIC CEng MIStructE

J W Rackham BSc MSc DIC PhD CEng MICE

A Way MEng CEng MICE

Published by:
The Steel Construction Institute
Silwood Park
Ascot
Berkshire SL5 7QN

Tel: 01344 623345
Fax: 01344 622944

© 2004 The Steel Construction Institute

Apart from any fair dealing for the purposes of research or private study or criticism or review, as permitted under the Copyright Designs and Patents Act, 1988, this publication may not be reproduced, stored or transmitted, in any form or by any means, without the prior permission in writing of the publishers, or in the case of reprographic reproduction only in accordance with the terms of the licences issued by the UK Copyright Licensing Agency, or in accordance with the terms of licences issued by the appropriate Reproduction Rights Organisation outside the UK.

Enquiries concerning reproduction outside the terms stated here should be sent to the publishers, The Steel Construction Institute, at the address given on the title page.

Although care has been taken to ensure, to the best of our knowledge, that all data and information contained herein are accurate to the extent that they relate to either matters of fact or accepted practice or matters of opinion at the time of publication, The Steel Construction Institute, the authors and the reviewers assume no responsibility for any errors in or misinterpretations of such data and/or information or any loss or damage arising from or related to their use.

Publications supplied to the Members of the Institute at a discount are not for resale by them.

Publication Number: SCI P334

ISBN 1 85942 151 2

British Library Cataloguing-in-Publication Data.

A catalogue record for this book is available from the British Library.

FOREWORD

This guide was prepared for two reasons. Firstly, the introduction of BS 5950-1:2000 highlighted the need to check the sway stability of all frames, including braced frames. Whilst this was required in the 1990 version of the Standard, common practice was to ignore the check for braced frames. Frame stability checks are therefore relatively new in practice, and deserve explanation. Secondly, no complete guide existed for what is a very important form of construction in the UK – multi-storey steel-framed buildings. Whilst the design of elements within a frame, such as beams and columns, is covered elsewhere, and is not of itself particularly challenging, a guide covering the wider aspects of multi-storey design was considered to be a valuable addition.

This guide does not attempt to repeat exhaustive coverage on the design of elements if these are covered elsewhere, but directs the reader to the resources that already exist. Instead, the guide offers advice on layout, floor systems and typical element sizes, so that the scheme design can be carried out with confidence.

The authors are indebted to their colleagues at The Steel Construction Institute for their input and advice. In addition, a number of other individuals have contributed to this guide and their input is gratefully acknowledged:

Mr W Brown	W. F. Brown Associates Limited
Mr J Brennan	Barrett Steel Buildings Limited
Mr T Hope	Arup

Additional review comments were received from Corus Construction & Industrial, Technical Sales and Marketing

The preparation of this guide was funded entirely by Corus Construction & Industrial, and their support is gratefully acknowledged.

Contents

		Page No.
FOREWORD		iii
SUMMARY		vii
1	INTRODUCTION	1
2	BUILDING DESIGN	2
	2.1 Design synthesis	2
	2.2 Ground conditions	2
	2.3 Site conditions	3
	2.4 Construction programme	3
	2.5 Basic layout	4
	2.6 Service integration	5
	2.7 Floor dynamics	6
	2.8 Fire safety	7
	2.9 Design life	8
	2.10 Acoustic performance	8
	2.11 Thermal performance	9
3	LOADING	10
	3.1 Types of loading	10
	3.2 Load combinations	12
4	FLOOR SYSTEMS	13
	4.1 Short-span composite beams and composite slabs with metal decking	14
	4.2 *Slimdek*	18
	4.3 Cellular composite beams with composite slab and steel decking	24
	4.4 *Slimflor* beams with precast concrete slabs	28
	4.5 Long-span composite beams and composite slabs with metal decking	32
	4.6 Composite beams with precast units	36
	4.7 Non-composite beams with precast units	40
	4.8 Beam connections	44
5	COLUMNS	48
	5.1 Column loading	48
	5.2 Column design	49
	5.3 Initial sizing	51
	5.4 Splices	51
	5.5 Splice design	52
	5.6 Column bases	55
	5.7 Bases to braced bays	56
6	BRACING SYSTEMS	59
	6.1 Introduction	59
	6.2 Horizontal diaphragms	59
	6.3 Horizontal bracing design	60
	6.4 Vertical bracing	62
	6.5 Vertical bracing design	63
	6.6 Bracing members and connections	64

7	**FRAME STABILITY**		**71**
	7.1 Introduction		71
	7.2 Frame behaviour		71
	7.3 Minimum lateral load		72
	7.4 Minimum wind load		72
	7.5 Frame stability checks		72
	7.6 Amplification to allow for sway-sensitivity		73
	7.7 Frame stability and ULS load combinations		74
8	**ROBUSTNESS**		**76**
	8.1 Introduction		76
	8.2 Structural integrity in BS 5950-1		76
	8.3 Avoiding disproportionate collapse		77
9	**FIRE RESISTANCE**		**82**
	9.1 Introduction		82
	9.2 Periods of fire resistance		84
	9.3 Fire protection systems		84
	9.4 Sources of further advice		85
10	**REFERENCES**		**86**
APPENDIX A	Demonstration of frame stability calculations		91
	A.1 Frame Stability Example		91

SUMMARY

This publication covers the design of braced, steel-framed multi-storey buildings, and offers guidance on the structural design of the superstructure. Recognising that 'building design' issues heavily influence the design of the superstructure, this publication also offers general advice on such issues as foundations, building layout, service integration and construction programme.

Detailed guidance is given on the application of the frame stability checks specified in BS 5950-1:2000, and how the Standard directs that any significant second-order effects may be allowed for.

Details are given for the common floor systems used in most multi-storey structures, providing typical framing layouts, typical member sizes and construction depths.

The application of the 'robustness rules' contained in BS 5950-1:2000, intended to ensure adequate tying and the avoidance of disproportionate collapse, is discussed in detail. As the UK Building Regulations are amended (expected in 2004), it is expected that the rules in the Standard will have much wider application.

Loading, element design and connection design have modest coverage in this document – the reader is directed to other authoritative guidance that already exists.

Dimensionnement des portiques multi-étagés contreventés

Résumé

Cette publication est consacrée au dimensionnement des portiques métalliques, contreventés, utilisés dans les immeubles multi-étagés et peut servir de guidance pour le dimensionnement de ces structures. Reconnaissant que la conception de l'immeuble influence fortement le dimensionnement de la superstructure, la publication donne aussi des informations concernant les fondations, l'agencement de l'immeuble, l'intégration des services et le programme de construction.

Une guidance détaillée est apportée à la vérification de la stabilité du portique, selon la BS 5950-1 :2000 et à la prise en compte des effets significatifs de second ordre.

Des détails sont fournis concernant les systèmes habituels de planchers utilisés dans les structures multi-étagées ainsi que pour les dispositions types de portiques, leurs dimensions et portées classiques.

L'application des « règles de robustesse » données dans la norme BS 5950-1 :2000, en vue d'éviter des ruptures disproportionnées, est analysée en détail. Comme les UK Building Regulations vont être modifiées (en 2004, probablement), ces règles devraient jouer un rôle plus important dans le futur.

Les charges, le dimensionnement des éléments structuraux et des assemblages ne sont guère pris en compte dans la publication – le lecteur est renvoyé à d'autres publications existantes traitant de ces sujets.

Berechnung von mehrgeschossigen, unverschieblichen Tragwerken

Zusammenfassung

Diese Publikation behandelt den Entwurf unverschieblicher, mehrgeschossiger Stahltragwerke und bietet eine Anleitung zur Berechnung des Überbaus. Erkennt man, dass Entwurfsfragen die Berechnung des Überbaus stark beeinflussen, bietet diese Publikation auch allgemeine Ratschläge zu Fragen wie Gründung, Grundriss, Installationsführung und Montage.

Ausführlich behandelt ist die Anwendung der Stabilitätsnachweise entsprechend BS 5950-1:2000 und die Berücksichtigung maßgebender Effekte der Theorie 2. Ordnung.

Details zu gebräuchlichen Deckensystemen, die bei der Mehrzahl mehrgeschossiger Gebäude verwendet werden, sind enthalten; sie zeigen typische Raster, Querschnittsgrößen und Konstruktionshöhen auf.

Die Anwendung der „Robustheitsregeln", die in BS 5950-1:2000 enthalten sind, wird im Detail besprochen; sie gewährleisten eine ausreichende Verbindung der Stützen mit den Decken und verhindern ein unverhältnismäßiges Versagen. Da die Bauvorschriften des Vereinigten Königreichs geändert werden (voraussichtlich in 2004), wird erwartet, dass die Regeln der Norm viel mehr Anwendung finden werden.

Lastannahmen und die Berechnung von Bauteilen und Verbindungen werden in diesem Dokument nur in bescheidenem Maße behandelt – der Leser wird auf die bereits bestehenden, maßgebenden Anleitungen verwiesen.

Proyecto de estructuras arriostradas de varias plantas

Resumen

Esta publicación abarca el proyecto de edificios de varias plantas con estructuras de acero arriostradas y da orientaciones sobre el dimensionamiento estructural de la superestructura. Aceptando que los temas de proyecto conceptual influyen fuertemente en el dimensionamiento la obra también contiene recomendaciones tales como las cimentaciones, la distribución e implantación, la integración de los servicios y el programa constructivo.

Se dan consejos muy detallados para la aplicación de los controles de estabilidad de la estructura especificadas en BS 5950-1:2000 y como la Norma establece que deben considerarse todos los efectos de segundo orden que sean importantes.

Se dan detalles respecto a los sistemas de forjado más habituales en las estructuras de varios pisos incluyendo distribuciones típicas, dimensiones habituales de las piezas detalladas y detalles de construcción.

Se analiza en detalle la aplicación de la "regla de robustez" contenida en BS 5950-1:2000 cuyo objetivo es garantizar un atado adecuado y un colapso desproporcionado.

Puesto que las UK Building Regulations serán corregidas (previsiblemente alrededor de 2004) se espera que las reglas de la Norma tenga un campo de aplicación más amplio.

Las cargas, el dimensionamiento de piezas y uniones tienen un tratamiento breve en este documento y el lector es dirigido a otras guías ya existentes.

Progettazione di telai multipiano controventati

Sommario

Questa pubblicazione tratta l'argomento degli edifici multipiano intelaiati in acciaio e con sistemi di controvento e costituisce una guida alla progettazione strutturale dell'ossatura portante. Come noto i requisiti associati alla progettazione della costruzione influenzano in modo incisivo anche aspetti legati alla struttura e pertanto vengono riportate informazioni di corredo sui sistemi di fondazione, sugli schemi funzionali dell'edificio, sull'integrazione degli impianti e sulla gestione del cantiere.

Vengono richiamate in dettaglio le principali regole per le verifiche di stabilità del telaio in accordo alla normativa BS 5950-1:2000, con particolare attenzione al caso in cui siano presenti significativi effetti del secondo ordine.

Sono considerati i più diffusi sistemi di impalcato per edifici multipiano, e vengono fornite indicazioni pratiche sulle tipologie, sulle dimensioni degli elementi e sugli ingombri.

Viene discusso in dettaglio come l'applicazione degli "affidabili criteri" riportati nella normativa BS 5950-1: 2000 porti indubbiamente a garantire adeguata sicurezza e ad evitare collassi pericolosi. Poiché le regolamentazioni del Regno Unito per gli edifici sono al momento in fase di emendamento (entro il 2004) sono attese regole progettuali caratterizzate da un campo di applicazione sicuramente più ampio.

Argomenti come carichi e progettazione di membrature e collegamenti sono marginalmente affrontati in questa pubblicazione ed è invece fatto rimando alla bibliografia disponibile.

Dimensionering av stagade flervåningsstommar

Sammanfattning

Denna publication omfattar dimensionering av stagade stålstommar till flervåningsbyggnader, och erbjuder vägledning för konstruktionsutformning av överbyggnaden. Med vetskapen att byggnadens utformning kraftigt påverkar överbyggnadens, erbjuder denna publikation även råd kring frågor som grundläggning, byggnadsutformning, integrering av installationer samt byggprogram.

Detaljerad vägledning ges för tillämpning av stomstabilitetskontrollerna enligt BS 5950-1:2000, och hur standarden medför att eventuella betydande andrahandseffekter kan tillåtas.

Detaljer ges för de vanliga bjäklagssystemen i de flesta typer av flervåningsbyggnader i form av typiska utformningar av stommen, typiska komponentstorlekar samt konstruktionshöjder.

Tillämpningen av 'robusthetsreglerna' i BS 5650-1:2000, med syftet att säkerställa tillfredsställande inspänning och undvika fortskridande ras, diskuteras i detalj. Eftersom de brittiska byggreglerna omarbetas (förväntas färdiga 2004) kan det förväntas att reglerna i standarden kommer att få en mycket bredare användning.

Laster samt dimensionering av element och förband har en blygsam täckning i detta dokument – läsaren hänvisas till annan existerande officiell vägledning och normer.

1 INTRODUCTION

Detailed guidance on the design of structural elements and connections within steel-framed buildings has been provided in numerous publications, but there has been little overall guidance on scheme design for braced multi-storey buildings in the UK. This publication sets out general guidance on scheme design for buildings designed to BS 5950 and indicates appropriate reference sources for detailed design considerations. One detailed design issue which has had scant discussion until now is the stability of braced frames. When BS 5950-1:2000[1] was issued in 2001, attention was drawn to the (apparently) new requirements to check frame stability, even for braced frames. In fact, the 1990 version of the Standard required a frame stability check for structures with rigid joints, and noted that sway stability could be provided by braced frames or joint rigidity, cores or shear walls.

Whilst the requirements within the Standard are therefore not entirely new, the earlier Standard did not explicitly cover stability issues in simple construction and common practice has generally been to ignore frame stability checks, particularly for braced frames.

Within this document, particular emphasis is placed on the checks of frame stability, but with a practical emphasis rather than on the underlying theory. Only the simple manual approaches to checking frame stability (and, if necessary, allowing for second-order effects) are covered in this document. The simple approach described in BS 5950-1:2000 has the beneficial effect of allowing the designer to retain a 'feel' for the behaviour of the structure, and is to be commended.

Section 8 of this guide covers the important topic of ensuring robust construction – by tying columns together and by other measures. This guidance is currently important where the structure is over five storeys, but is likely to have increased applicability when the Building Regulations[2] are amended.

Section 4 covers floor solutions and is intended to assist the designer choose an appropriate solution at the scheme design stage. The choice of floor solution will generally be related to the permitted column layout and to the degree of service integration, perhaps driven by a need to minimise construction depth. Section 4 describes how services are incorporated into each solution, describes typical layouts and identifies the critical design checks in the subsequent numerical check of the elements.

2 BUILDING DESIGN

2.1 Design synthesis

In most buildings, the superstructure design, whilst important, is of much less priority than defining the functional aspects of the building. The structural configuration is strongly influenced by issues such as the clear floor spaces, the vertical circulation, the ventilation and the lighting. In addition, ground conditions often have a major influence on the design solution, and may dictate the column layout. Speed of construction and minimum storage of materials on site may be critical, and the Main Contractor's preferences (or aversions) to a particular form of construction are also important.

The cost of the building superstructure is generally only 10% of the total capital cost – foundations, services and cladding are often more significant. The design of the superstructure cannot be completed in isolation – in reality the building design must be resolved before the structural frame can be completed. This Section offers outline guidance on the issues likely to affect the scheme design of the frame. More details can be found in the references.

The British Council for Offices (BCO) guide *Best practice in the specification for offices*[3] is an excellent summary of design issues to be considered in any structure, and is recommended reading. The BCO guide covers planning issues, key design parameters, performance criteria and completion, with many recommendations on best practice.

2.2 Ground conditions

The ground conditions may dominate the possible column layout. Increasingly, structures must be constructed on poor ground conditions, or on 'brownfield' sites, where earlier activities have left a permanent legacy. It is often said that whilst the cost of a superstructure is relatively fixed, the foundation design can make a major difference to the cost of the scheme.

In city centres, major services and underground works, such as sewers and tunnels, are a major design consideration, often dominating the chosen solution.

Generally, poor ground conditions tend to produce a solution involving fewer, more heavily loaded foundations. This would necessitate longer spans for the superstructure. Many long span steel solutions are available[4]. Common solutions make use of cellular beams or fabricated beams, as described in Section 4.

Good ground conditions usually permit increased numbers of lightly loaded columns, and a shorter grid. Shorter spans permit the use of shallower beams, with the potential for a reduced construction depth, or for uninterrupted soffits.

2.3 Site conditions

A confined site can place particular constraints on the structural scheme. Site constrains may limit the physical size of the elements that can be delivered and erected, leading to shorter column lengths between splices, and precluding long-span beams. On a constrained site, composite flooring may be the preferred floor solution compared to precast units, as the decking may be delivered in short lengths, needing only a small crane. On a congested site, to have steel deliveries, precast unit deliveries and a crane on site at the same time may prove impossible.

On very congested sites, access may demand that steel is erected directly from a delivery lorry in the road. This may preclude working at certain times in the day, or require working over the weekend, making the erection programme relatively inflexible. Erection directly from a delivery lorry is likely to favour simple, shorter components.

Smaller inner-city sites are often served by a single tower crane, which is used by all trades. In these circumstances, craneage is limited, and smaller piece counts are an advantage.

2.4 Construction programme

The construction programme will be a key concern in any project, and will need to be considered at the same time as considering the cost of structure, the services, cladding and finishes. As the structural scheme will have a key influence on both programme and cost, a solution cannot be reached in isolation. The shortest programme is undoubtedly required, which will necessitate full integration of following trades, usually whilst the steel is being erected. Structural solutions which can be erected safely, quickly and allow early access for the following trades are required.

Erection rates are dominated by "hook time" – the time connected to the crane. Fewer pieces to erect, or more cranes, will reduce the erection programme.

Cranes

The number of cranes on a project will be dominated by

- The site footprint – can more cranes be physically used?

- The size of the project – can more than one crane be utilised, or is the structure too small?

- Commercial decisions on cost and programme benefits.

Multi-storey structures are often erected using a tower crane. As tall buildings are erected, the increased time lifting the item into position from ground level is noticeable. More significantly, there are usually competing demands from other trades for the use of tower cranes, which can slow overall progress for the steelwork erection. For larger projects, erection schemes that enable other trades to commence their activities in an integrated way as the steelwork progresses will be required. This may impact, for example, the choice of floor solutions.

Composite floors

Composite floors involve the laying out of profiled steel decking, which is lifted onto the steelwork in bundles and usually man-handled into position. Safety nets are erected immediately after the steelwork and before the decking operation. Steelwork already erected at upper levels does not prevent decking being lifted and placed, although decking is usually placed as the steelwork is erected. Completed floors may be used as a safe working platform for subsequent erection of steelwork, and allow other works to proceed at lower levels. For this reason, the upper floor in any group of floors (usually three floor levels) is often concreted first, bringing forward the time when the floor has cured.

Precast concrete planks

Placing of precast concrete planks becomes difficult if the planks must be lowered through erected steelwork. Better practice is to place the planks as the steelwork for each floor is erected, and to have the plank supply and installation as part of the Steelwork Contractor's package is often an advantage. The Steelwork Contractor can arrange material delivery to suit his own erection method. Generally, columns and floor steelwork will be erected, with minimal steelwork at upper levels – enough to stabilise the columns, until the planks have been positioned. Steelwork for the upper floors will then continue.

Erection rates

As an indication only, an erection rate of between 20 and 30 pieces per day is a reasonable rate. With average piece weights, this equates to approximately 10 tonnes per day.

In 1990, a case study was prepared of the erection of Senator House, an eight-storey office block, generally erected with one tower crane. The erection rates achieved on this project were 195 pieces per week, and 101 tonnes per week. A special emphasis was placed on minimising the time spent slewing and hoisting.

2.5 Basic layout

The choice of the basic building shape is usually the Architect's responsibility, constrained by the client and such issues as the site, access, building orientation, parking, landscaping and local planning requirements. The following general guidance affecting the structure itself is taken from the BCO specification.

- Building plan depth should be between 13.5 and 21m.
- Naturally lit and ventilated zones extend a distance of twice the floor-to-ceiling height from the outer walls – artificial light and ventilation will be required elsewhere.
- Four storeys are optimum for cost efficiency and floor plate efficiency.
- Column grids of 7.5 m to 9 m are economic.

The BCO guide notes that atria improve floor plate efficiency and because exposure to external climate is reduced, reduce the capital cost of the envelope and running costs. Atria make a significant contribution to the effectiveness of the office environment and amenity.

2.6 Service integration

Most large office type structures will need significant air conditioning, which will require both horizontal and vertical distribution systems. Comprehensive guidance is given in Reference [5], and a guide to service integration in Reference [6]. The provision for such systems is of critical importance for the superstructure layout, affecting the layout and type of members chosen.

The basic decision either to integrate the ductwork within the structural depth or to simply suspend the ductwork at a lower level affects the choice of member, the fire protection system, the cladding (cost and programme) and overall building height. Integrated services do not automatically need to be below the floor (i.e. in the ceiling void). Certain systems provide conditioned air from under a raised floor.

The most commonly used systems are the Variable Air Volume system (VAV) and the Fan Coil system. VAV systems are often used in buildings with single owner occupiers, because of their lower running costs. Fan Coil systems are often used in speculative buildings because of their lower capital costs.

Spatial aspects of vertical and horizontal service distribution are reviewed in Reference [5]. Generally, a zone of 450 mm will permit services to be suspended below the structure. An additional 150–200 mm is usually allowed for deflection, fire protection, ceiling and lighting units. Terminal units (Fan coil or VAV units) are located between the beams.

Service integration is achieved by passing services through penetrations in the supporting steelwork. These may be individual holes formed in ordinary steel beams, or multiple regular or irregular holes created by fabricating beams. Fabricated beams with regular circular cells (known as a cellular beam) are created by welding together two 'halves' of a rolled section. The top and bottom halves may be of different sizes and from different beams. Fabricated plate girders are created from flange and web plates, with a wide range of sizes and hole combinations.

The shallowest integrated floor solution is achieved with deep decking and special asymmetric beams, where services can be located in the troughs in the decking, and pass through the supporting steelwork, as shown in Figure 2.1. The size of the services is obviously limited in this arrangement.

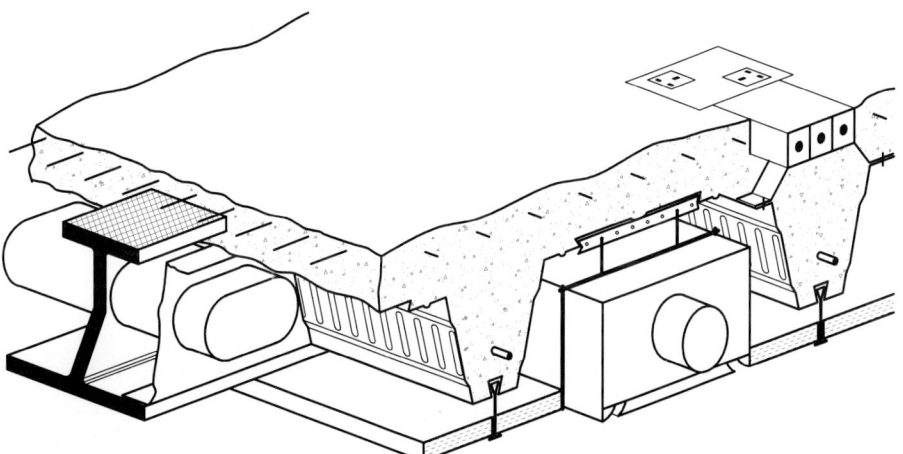

Figure 2.1 *Integration of services within Slimdek*

If there are no overall height constraints, it is usually cheaper to accommodate services below the floor structure. This obviously simplifies the layout and eases any subsequent replacement. The penalty is an increased construction depth of each floor, and increased cladding areas around the structure. Both the increased cost of cladding and the possible programme implications should be considered, as, for example, a reduction in several brick courses at each floor could produce benefits in time and cost.

The BCO specification[3] encourages integration, noting that significant savings in overall storey height can be obtained by co-ordinating structure and services. The BCO specification also recommends that integration should not be pursued to such extremes that buildability, access to services and flexibility for modification are compromised.

2.7 Floor dynamics

Until recently, floor response was assessed by calculating the fundamental frequency of the floor. If the fundamental frequency was greater than 4Hz, the floor was considered to be satisfactory. Whilst this was generally acceptable for busy workplaces, it is not appropriate for quieter areas of buildings where vibrations are more perceptible.

A more appropriate approach is an assessment based on a 'response factor' that takes into account the amplitude of the vibration, which is normally measured in terms of acceleration. Higher response factors indicate increasingly dynamic floors – more noticeable to the occupants. Comprehensive guidance is contained in References [7] and [8], with recommended limiting response factors for different office environments.

In practice, response factors are reduced (i.e. vibration is less noticeable) by increasing the mass participating in the motion. Long-span beams are generally less of a dynamic problem than shorter spans, which is quite contrary to perceived wisdom based on frequency alone.

Beam layout is often important, as longer continuous lines of secondary beams in composite construction result in lower response factors than shorter lengths, because more mass participates in the motion with longer lines of beams. Figure 2.2 shows two possible arrangements of beams. In both cases, 'primary' beams support 'secondary' beams and 'secondary' beams support the composite slab. The response factor for arrangement (b) will be lower (less noticeable to occupants) than arrangement (a), as the participating mass is increased in arrangement (b).

Damping reduces the dynamic response of a floor. Floor response is decreased by partitions at right angles to the main vibrating elements (usually the secondary beams). Bare floors during construction are likely to feel more 'lively' than when occupied.

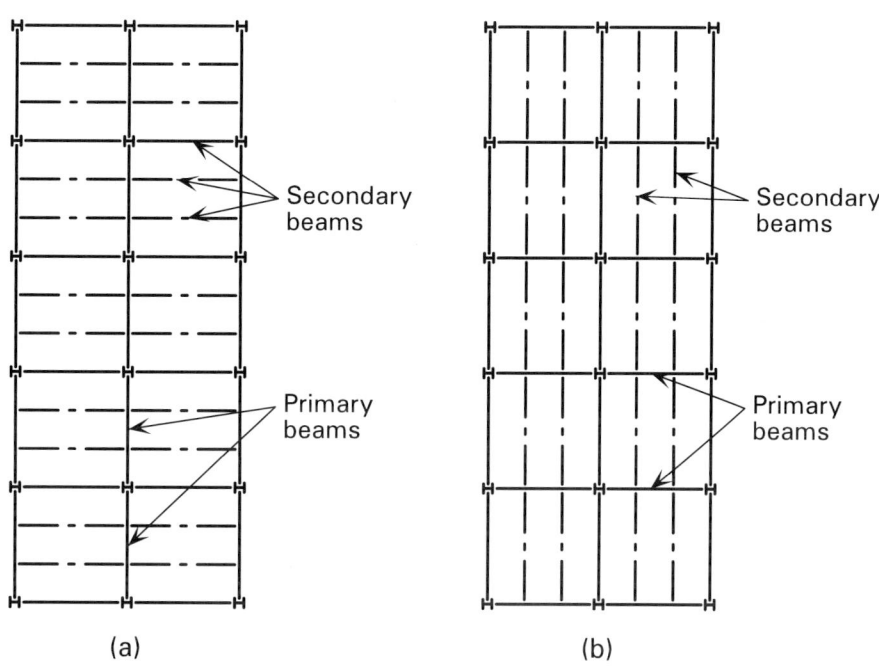

Figure 2.2 *Alternative beam layouts*

2.8 Fire safety

Building designers will need to consider the effects of fire when arranging the building layout, and when choosing the structural configuration. Whilst not affecting structural design, building designers will address such issues as:

- Means of escape.
- Size of compartment.
- Access and facilities for the Fire Service.
- Limiting the spread of fire.
- Smoke control and evacuation.
- The adoption of sprinklers.

Fire resistance

In addition to the above, structural performance in the event of a fire must meet prescribed standards, expressed as a period of fire resistance of the structural components. Fire resistance of steel components is discussed in more detail in Section 9. As an alternative, a 'fire engineering' approach may be followed which accounts for the fire safety of the whole building, considering the structure use, the hazards, the risks and how these are addressed. A 'fire engineering' approach is beyond the scope of this document.

In general, the structural engineer will wish to:

- Adopt schemes with fewer beams to fire protect.
- Investigate all opportunities to leave steel unprotected.
- Acknowledge that service integration may complicate fire protection, and choose appropriate solutions such as intumescent coatings on cellular and other fabricated members.

- Acknowledge that site-applied protection may have programme implications, particularly if spray applied.

- Account for the final finish when choosing fire protection.

Structural fire safety: A handbook for architects and engineers[9] is recommended reading on this subject.

2.9 Design life

When proposing any structural scheme, it should be acknowledged that the structure itself will have a design life many times greater than other building components. For example, service installations have a design life of around 15 years, compared to a design life of around 50 years for the structure. Building envelopes for typical office construction have a design life of between 30 and 50 years. The implications for the structural solutions can be profound – recognising that a solution that facilitates easy replacement or upgrading of the services reduces the whole life costs of the structure considerably.

Similarly, the space usage of the interior is likely to change constantly. Schemes that allow maximum flexibility of layout are to be preferred. The BCO specification recommends that the structure be designed for flexibility and adaptability, achieved with:

- Longer floor spans.

- Higher ceilings.

- Ease of maintenance.

The BCO specification recommends that the structure be designed to allow as many servicing and layout options as possible, with a clear strategy for flexibility and future adaptability of the structure.

2.10 Acoustic performance

Residential structures

Acoustic performance of residential structures is covered by Parts E1 to E3 of the Building Regulations[10].

Part E1 considers protection against sound from other parts of the building and gives specific performance requirements for separating walls and floors. The requirements cover both airborne sound and, for floors, impact sound transmission.

Part E2 covers sound within a dwelling, and requires that such elements as internal walls around bedrooms must provide reasonable resistance to sound transmission.

The requirements for Part E1 can be appreciated by considering the 'Robust Standard Details' (RSD) that have been developed. The RSD are systems and details that have been demonstrated by in-situ testing to exceed the standards specified in the Building Regulations, and may be used in Domestic construction. If the RSD are not used in domestic construction, compliance with the Regulations must be demonstrated by post-completion testing.

The RSD may also be used with confidence in other forms of construction, although testing would still be required.

Office buildings

The BCO specification recommends criteria for residual noise, after accounting for attenuation by the building façade, suggesting limits for open plan offices, cellular offices and conference rooms. Criteria are also given for the acceptable noise from building services in the same categories of office.

BS 8233

BS 8233[11] contains maximum and minimum ambient noise level targets for spaces within buildings. These are appropriate for comfort in both commercial premises and residential accommodation. The Standard also includes acoustic information on noise from traffic, aircraft and railways.

Structural implications of acoustic performance standards

To meet acoustic performance standards, the construction details will need special attention, particularly where walls meet floors and ceilings (known as flanking details). As a minimum, the structural designer needs to be aware of the detailing required to meet the acoustic performance standards when considering structural options. Whilst the basic structure may not be affected, floating floors and suspended ceilings may be required, which will impact any decision on service integration. Separating walls meeting the requirements of Part E of the Building Regulations are likely to be of twin skin construction, facilitating the use of bracing within the wall construction.

Further guidance on the acoustic performance of structural systems can be found in References [12], [13] and [14].

2.11 Thermal performance

Thermal performance of structures (other than Dwellings) is covered by Part L2 of the Building Regulations[15]. Apartments are covered by Part L1. There are three approaches that mey be used to demonstrate that the Regulations have been met – the most common approach is the Elemental Method, which requires the designer to demonstrate by calculation that the specified details for the walls and roof meet the specified insulation values. The Regulations also specify that there should be no significant thermal bridges or gaps in the insulation, and for buildings with over 1000 m^2 of floor area, specify that airtightness must be demonstrated by physical testing.

Whilst these issues may appear to be traditionally the Architect's responsibility, the structural engineer must be intimately involved in the development of appropriate details and layout. Steel beams may have to be placed in non-preferred locations so that they can be insulated. This may introduce eccentricity into the structure, affecting the design of the member and its connections. Similarly, supporting systems for cladding may be more involved, again involving eccentric connection to the supporting steelwork.

Steel members that penetrate the insulation, such as balcony supports, need special consideration and detailing to avoid thermal bridges. Thermal bridges not only lead to heat loss, but may also lead to the formation of condensation on the inside of the building, with the potential of corrosion of the steelwork and damage to internal fittings.

3 LOADING

3.1 Types of loading

There are three principal types of loading that need to be considered in the design of multi-storey framed buildings:

- Dead loading.
- Imposed loading.
- Wind loading.

3.1.1 Dead loading

Dead loading is permanent, stationary loading and consists of the self-weight of the structure. This should include the weight of the structural frame, the floor system, the services, the ceiling and the finishes. Table 3.1 gives typical weights for common elements.

Table 3.1 *Typical weights for building elements*

Element	Typical weight
Precast units (Spanning 6 m, designed for a 5 kN/m^2 imposed load)	3 to 4.5 kN/m^2
Composite slab (Normal weight concrete, 130 mm thick)	2.6 to 3.2 kN/m^2
Composite slab (Light weight concrete, 130 mm thick)	2.1 to 2.5 kN/m^2
Services	0.25 kN/m^2
Ceilings	0.1 kN/m^2
Steelwork (low rise 2 to 6 storeys)	35 to 50 kg/m^2
Steelwork (medium rise 7 to 12 storeys)	40 to 70 kg/m^2

3.1.2 Imposed loading

Imposed loading is the non-permanent loading that is likely to be applied to the structure during its life (other than wind load). Imposed loads should include gravity loads due to occupants, equipment, furniture, movable partitions, stored materials and snow. The magnitude of the imposed loading will vary according to building use and the use of any specific floor (or roof) area being considered – different values are applied for a plant room or storage area, for example.

Imposed loading on floors

BS 6399-1[16] prescribes minimum imposed floor loads for different building uses. A commonly used imposed loading for a commercial office is 4 kN/m^2 plus 1 kN/m^2, often known as '4 plus 1'. The additional 1 kN/m^2 is to allow for movable partitions. Some designers use 5 kN/m^2 or even '5 plus 1'.

Table 3.2, extracted from BS 6399-1 gives some typical minimum imposed floor loading values. The concentrated loads are applied independently from the distributed loads, and are used to check punching or crushing. For concentrated loads, the Standard suggests that a realistic area of application is used, or, in the absence of any other data, a square area 50 mm by 50 mm. The concentrated loads may also be applied to members at any location, to produce bending moments and shears. In orthodox construction, it is highly unlikely that the concentrated loads affect the design; common practice is to ignore them.

Table 3.2 *Minimum imposed loads (from BS 6399-1)*

Type of activity	Specific use	Uniformly distributed load kN/m²	Concentrated load kN
Domestic and residential	Bedrooms in hotels and motels. Hospital wards. Toilet area.	2.0	1.8
Offices	Offices for general use	2.5	2.7
Areas with tables	Classrooms	3.0	2.7
Areas with possible physical activities	Dance halls and studios, gymnasia, stages	5.0	3.6
Shopping areas	Shop floors for the sale and display of merchandise	4.0	3.6
Warehousing and storage areas	General storage (Some particular storage types are specified in BS 6399-1)	2.4 per metre of storage height	7.0
Areas for plant and equipment	Plant rooms, boiler rooms, etc (including weight of machinery)	7.5	4.5

Alternative loading values

An often-used alternative to the loadings specified in BS 6399-1 are the loadings recommended by The British Council for Offices[3] (BCO), who recommend the following floor loadings.

Standard allowances for live load:

General area: 2.5 kN/m² over approx 95% of each potentially sub-lettable floor area.

High loading area: 7.5 kN/m² over approx 5% of each potentially sub-lettable floor area and not in primary circulation routes.

Standard allowances for dead load:

Demountable partitions: 1.0 kN/m²

Raised floors, ceiling and building services equipment: 0.85 kN/m²

The BCO take issue with the imposed loading specified in BS 6399-1. The following is taken from their guidance:

Historically, UK office buildings have been designed and marketed with floor loadings significantly higher than the current British Standard loading threshold of 2.5 kN/m². Research has shown this to be an over-provision.

It should be borne in mind that this is just one industry viewpoint. It has been suggested by others that the imposed floor loading for the general office area category of 2.5 kN/m² is too low, particularly for situations where the floor usage may change, which may result from a change in the building occupier. Often, a conservative view is taken and the higher imposed load chosen as the design load.

In view of the forgoing, it is obviously important to clarify the required floor loading with the client at the earliest stage, and to record this information in the Health and Safety File.

Imposed loads on roofs

BS 6399-3[17] specifies an absolute minimum imposed roof load of 0.6 kN/m^2. This figure may be exceeded at high altitude, and in the North of the UK, where greater snow load is experienced. BS 6399-3 must be consulted and the imposed roof load calculated for the actual site location.

BS 6399-3 also identifies where local drifts may form from redistributed snow, for example in valleys or against obstructions. This accidental load case must also be considered, although the partial load factor (see Section 3.2) is reduced to 1.05.

BS 6399-3 identifies concentrated loads to apply to the roof, over a specified contact area. The concentrated loads usually have no impact on the design of the main structural elements.

3.1.3 Wind loading

Wind loading is the load applied to the structure due to the effects of wind pressure and suction. Wind loading should be determined using BS 6399-2. Guidance on applying BS 6399-2 is provided in SCI online publication ED001 *Recommended application of BS 6399-2*[18]. This guidance sets out a recommended procedure for applying the provisions of the Standard, particularly for those designers using the Standard for the first time. When calculating overall loads (for example to design the bracing), the overall force coefficients found in Table 5a of BS 6399-2 should be used.

Software is available, as stand-alone commercial packages, and as part of software available from cold rolled component manufacturers that is used for the design of side rails and purlins.

3.2 Load combinations

For the ultimate limit state, loads should be multiplied by the appropriate load factors and combined with other loads to form realistic load combinations. The common load combinations and corresponding load factors are given by BS 5950 and are shown in Table 3.3. The application of the notional horizontal forces shown in Table 3.3 is discussed in detail in Section 7.

Table 3.3 *Partial factors for loads*

Load combination	Dead	Imposed	Wind	Notional horizontal forces*
Dead + Imposed	1.4	1.6		1.0
Dead + Imposed + Wind	1.2	1.2	1.2	
Dead + Wind	1.4		1.4	
Dead + Wind (Overturning)	1.0		1.4	

* Calculated as 0.5% of [factored dead plus factored imposed] loads.

For the serviceability limit state, unfactored loads should be used.

4 FLOOR SYSTEMS

In addition to their obvious load-carrying function, structural floors often act as horizontal diaphragms, ensuring horizontal loads are carried to the vertical bracing. Floor components (the floor slab, deck units and the beams) will also require a certain fire resistance, as described in Section 9. Services may be integrated with the floor construction, or like the ceiling, simply suspended below the floor. Structural floors may have a directly-fixed floor finish, or may have a screed, or a raised secondary floor above the structure. Raised floors allow services (particularly electrical and communication services) to be distributed easily around highly serviced accommodation.

This Section describes seven floor systems often used in multi-storey buildings. The main characteristics of each floor system are described, with guidance on important design issues.

This Section does not contain detailed design procedures but directs the reader to the sources of design guidance, manufacturer's literature or software.

The following floor systems are covered:

- Short-span composite beams and composite slabs with metal decking.
- *Slimdek*.
- Cellular composite beams with composite slabs and steel decking.
- *Slimflor* beams with precast concrete units.
- Long-span composite beams and composite slabs with metal decking.
- Composite beams with precast concrete units.
- Non-composite beams with precast concrete units.

Composite construction

In the following Sections, design approaches are suggested for composite construction, where decking must be chosen and a type of concrete assumed.

Decking may have a re-entrant or trapezoidal profile. Re-entrant decking uses more concrete than trapezoidal decking, but has increased fire resistance for a given slab depth. Trapezoidal decking generally spans further than re-entrant decking, but the shear stud resistance is less with trapezoidal decking than with re-entrant decking.

Generally, lightweight concrete (LWC) is proposed in this document, unless a directly-bonded floor is specified. Designers should note that LWC is usually more expensive than normal weight concrete (NWC), and may not be available in all areas of the country. Ideally, the choice of concrete should be made in conjunction with the Main Contractor, in order to produce an optimum scheme.

4.1 Short-span composite beams and composite slabs with metal decking

Description This floor system consists of downstand steel beams with shear connectors welded to the top flange to enable the beam to act compositely with an in-situ composite floor slab.

Framing arrangements normally involve the slab spanning 3 to 4 m to secondary beams, which are in turn supported by primary beams. Secondary and primary beams are usually composite. Edge beams are often non-composite.

The floor slab comprises a shallow ribbed metal decking and a concrete topping, which act together compositely. Slabs are typically 130 mm thick and the decking about 60 mm deep in galvanized strip, with a material thickness of 0.9 to 1.2 mm.

The shear connectors are normally site-welded through the decking to provide a strong fixing to the beam, and to enable the decking to provide restraint to the beam during the construction stage when the concrete is being poured.

Mesh reinforcement, normally A142 or A193, is placed in the slab to enhance the fire resistance of the slab, to help spread localised loads, to act as shear reinforcement around the shear connectors and to reduce cracking in the slab.

The decking is normally designed to support the wet weight of the concrete and construction loading as a continuous member over at least two spans, but the composite slab is normally designed as simply supported between beams.

Typical beam span range Secondary beams: 6.0 m to 7.5 m at 3 to 4 m spacing.

Primary beams: 6.0 m to 9.0 m.

Main design considerations for the floor layout Secondary beams should be spaced closely enough to avoid propping the decking, as propping can be expensive and disruptive on site.

Services will need to pass under beams, and thus affect the overall floor zone.

Overall floor zone may be governed by depth of edge beams. They may need to be deeper than internal beams because of more onerous serviceability criteria in supporting the cladding. Also, the use of non-composite edge beams avoids the need for detailing special U-bars around the shear connectors, but the beams may be deeper than a composite member.

Advantages Shallower beams than non-composite construction, lightweight, economic.

Disadvantages More columns needed than with long-span systems.

Deeper overall floor zone than shallow floor systems.

Beams require full fire protection for 60 min fire resistance and greater.

Services integration Main heating and ventilation units can be positioned between beams, but ducts must pass below beams. Small services may be taken through discrete holes in the web up to 150 mm diameter, where beam strength will allow.

Governing design criteria for beams Total deflections will usually govern for S355 secondary beams. Strength will usually govern for all S275 beams and for all primary beams.

When serviceability criteria govern, consider S275 sections, which are cheaper than S355 sections.

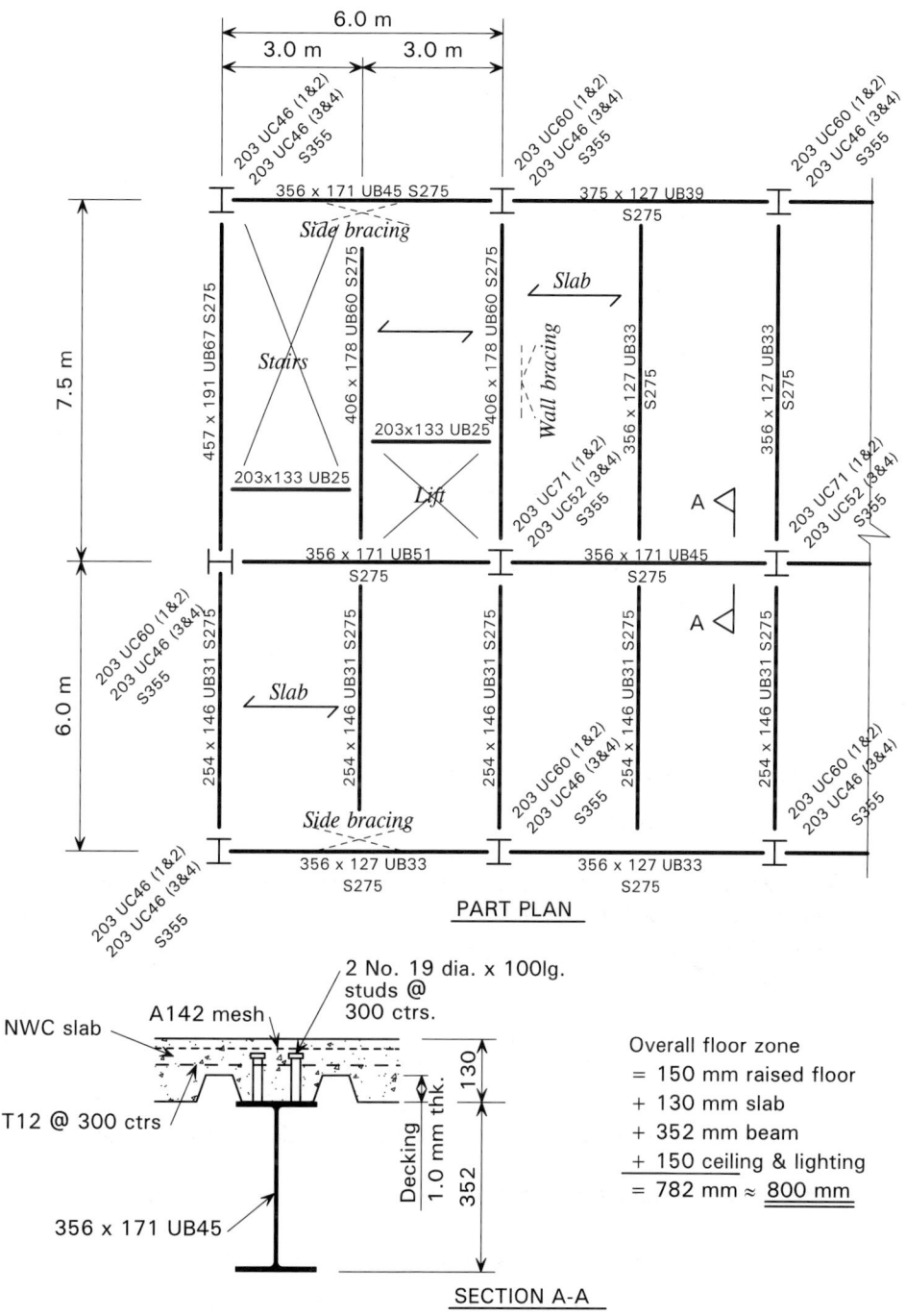

Figure 4.1 *Short-span composite beam ~ example of floor steelwork arrangement for 4-storey rectangular plan building*

Governing design criteria for decking/slab
Strength or deflection of the decking in the construction condition.

Fire resistance (affects concrete cover to the decking and mesh reinforcement size).

Strength or deflection in the composite condition.

Design approach
1. Assume secondary beams at 3 – 3.75 m spacing, on a 6 m, 7.5 m or 9 m grid

2. Choose decking and slab, using decking manufacturer's load tables or software. Assume LWC, unless there is a directly-bonded floor covering. Assume C35 concrete, and unpropped decking during construction. Ensure chosen slab and reinforcement meet the fire resistance required.

3. Design beams using software. Try studs at approximately 300 mm spacing for secondary beams (to suit trough spacings), and at 150 mm spacing on primary beams. Note that the orientation of the decking will differ between secondary and primary beams.

Typical section sizes
Composite beam depth (steel beam plus slab) ≈ span/16 to span/18
254 × 146 UB31 S275 for 6 m at 3.0 m spacing (secondary beam)
305 × 165 UB40 S355 for 7.5 m at 3.75 m spacing (secondary beam)
356 × 171 UB57 S355 for 7.5 m at 7.5 m spacing. (primary beam)
Usually one serial size deeper or one weight heavier for edge beams

Grade of steel
Secondary beam and edge beams: Usually S275.

Primary beam: Either S275 or S355.

Overall floor zone
Typically, 1200 mm for 7.5 m grid with 150 mm raised floor and air conditioning. Typically 700 mm for a 6m grid without services.

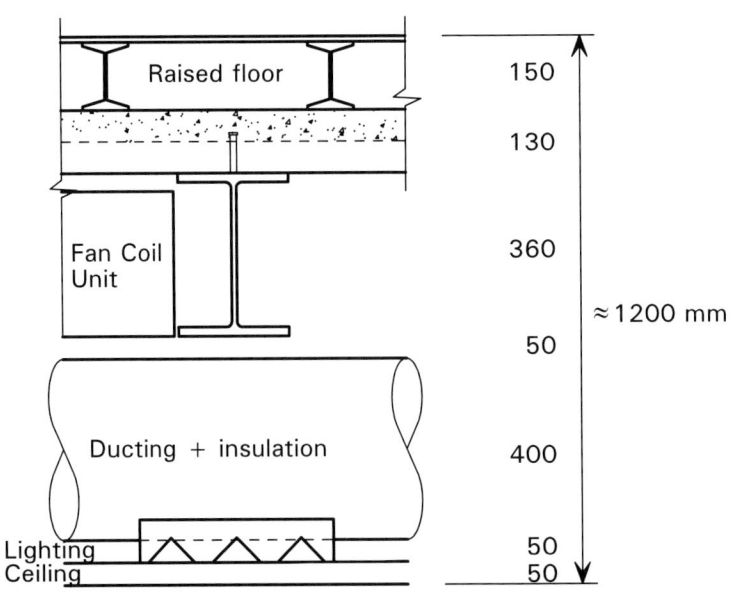

Figure 4.2 *Overall floor zone ~ typical short-span composite construction*

Type of concrete	Either normal weight concrete (NWC), 2350 kg/m^3 dry density, or lightweight concrete (LWC), 1850 kg/m^3 dry density, can be used.
	NWC has better sound reduction, so is better for residential buildings, hospitals, etc.
	LWC is better for overall building weight/foundation design, better span capability of slab, and has better fire insulating properties, enabling thinner slabs (10 mm less) to be used. It is not available in all parts of the UK. LWC is not considered suitable for directly-bonded floor coverings.
Grade of concrete	Use C30 as a minimum. Use C40 if concrete is to be used as a wearing surface.

Fire protection	Beams (typically):
	   Either Intumescent coating up to 1.5 mm thick for up to 90 minutes , or
	   Board 15 - 25 mm thick for up to 90 minutes
	   Note: P288[19] describes how beams may be left unprotected in certain areas.
	Columns (typically):
	   Board 15 mm thick for up to 60 minutes
	   Board 25 mm thick for 90 minutes

Connections	Simple (non-moment resisting) connections: double angle cleats, partial depth flexible endplates or finplates.
Design guidance	For choice of decking and composite slab design (including fire resistance); manufacturer's design tables. For best practice advice in design and construction; P300[20] For design charts and worked example for decking and beams; P055[21] For fire protection; the 'Yellow book'[22]

Software	Slab design - Comdek software, available from www.corusconstruction.com
	    - Deckspan software, available from www.rlsd.com/
	    - Multideck software, available from www.kingspanmetlcon.com/services/software/index.htm
	Beam design: BDES software, available from www.corusconstruction.com

4.2 Slimdek

Description

Slimdek is a shallow floor system comprising asymmetric floor beams (ASBs) supporting heavily ribbed composite slabs with 225 mm deep decking. ASBs are proprietary beams with a wider bottom flange than top. The section has embossments rolled into the top flange and acts compositely with the floor slab without the need for additional shear connectors. The decking spans between the bottom flanges of the beams and acts as permanent formwork to support the slab and other loads during construction. The in-situ concrete acts compositely with the decking and encases the beams so that they lie within the slab depth – apart from the exposed bottom flange.

Span arrangements are normally within a 6-9 m grid, with a slab depth of 280-350 mm. Decking requires propping at the construction stage for spans beyond about 6 m. Reinforcing bars (16–25 mm dia) need to be included in the ribs of the slab to give sufficient strength in the fire condition. The reinforcing bars also improve the composite floor strength in the normal condition.

Edge beams can be RHS *Slimflor* beams, which comprise a rectangular or square rolled hollow section with a flange plate welded underneath, ASBs or downstand beams. Ties, normally Structural Tees with the leg cast in the slab, are used to restrain the columns internally in the direction at right angles to the main beams.

A range of ASB sections is available in each of two serial sizes of 280 and 300 mm depth. Actual depths vary between 272 mm and 342 mm. Within this range there are five ASBs with relatively thin webs and five ASB(FE) (fire-engineered) sections with relatively thick webs (equal to or thicker than the flanges). The ASB(FE) sections offer a fire resistance of 60 minutes without additional protection in this form of construction with normal office loading. All ASBs are rolled in S355 steel.

Mesh reinforcement (A142 for 60 minutes fire resistance and A193 for 90 minutes) is placed in the slab over the ASB. If the top flange of the ASB is flush with the surface of the concrete, the slabs each side of the ASB will require tying together to meet robustness requirements, normally by reinforcement (typically T12 @ 600 ctrs) taken through the web of the ASB. ASBs are normally designed as non-composite if the concrete cover over the top flange is less than 30 mm. Note that a cover to the ASB of either zero or at least 30 mm is recommended (the aggregate/reinforcement cannot be accommodated easily in less than 30 mm depth).

Typical beam span range

6–7.5 m grids, typically, although 9 × 9 m possible.

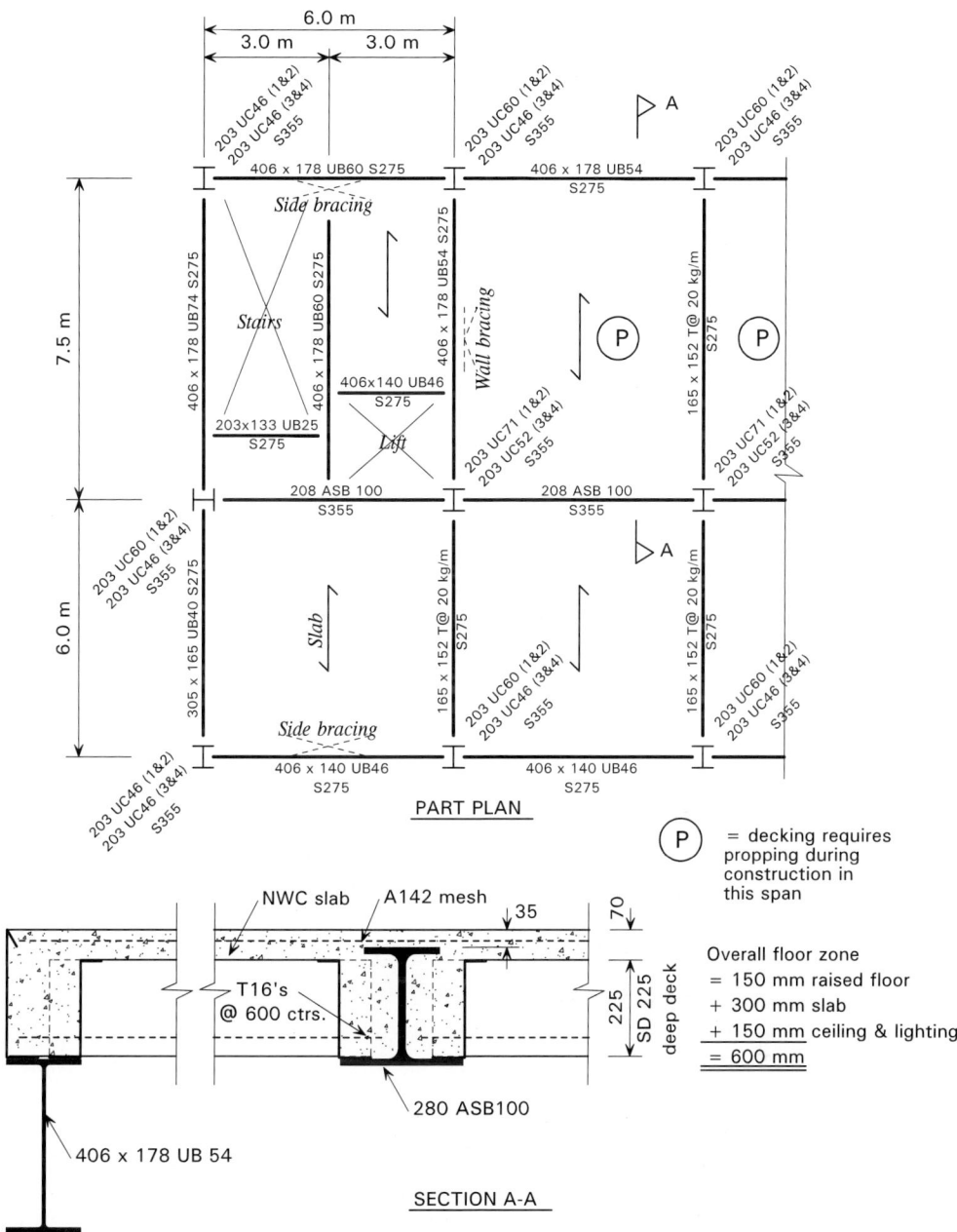

Figure 4.3 *Slimdek – floor steelwork arrangement for a four-storey rectangular building (central spine ASB and downstand edge beams)*

Main design considerations for the floor layout

A central spine of ASBs with decking spanning onto edge beams will generally be more economic than a series of parallel transverse ASBs, for buildings with a rectangular plan shape. Torsion may govern beam design at a change in direction of floor span and for edge beams. RHS *Slimflor* edge beams designed to resist torsional loading are likely to be the deepest member in a *Slimdek* floor.

Decking requires propping for spans over 6 m (propped twice at 9 m span).

Slab depth is influenced by the concrete cover to the deck (mainly for fire resistance), cover to the ASB (30 mm minimum, or zero), and cover to the edge beam. ASBs are designed as non-composite if the cover is less than 30 mm.

Detailing of connections around columns should be considered, as the ASB flanges are wider than the column and may need notching.

Advantages

Shallow floor zone – reduction in overall building height and cladding. Virtually flat soffit allows easy service installation and offers flexibility of internal wall positions.

Disadvantages

Steel weight is often heavier than other floor systems.

Connections require careful detailing due to the width of the bottom flange.

Services integration

Virtually flat soffit allows unrestricted access for services below the floor. Small services and ducts (up to 160 mm dia) can be passed through holes in the beam webs and between troughs in the decking.

Governing design criteria

Slab depth may be controlled by fire resistance, ultimate strength or concrete cover to ASB/edge beams

Deflections, fire resistance, strength or torsional loading governs the size of ASBs.

Design approach

1. Assume beams on a 6 m, 7.5 m or 9 m grid. (Note that decking over 6 m requires propping, which may affect the construction programme.)

2. Choose the decking and design the slab using software. Assume LWC unless there is a directly-bonded floor covering. Assume C35 concrete, with propping if required. Ensure chosen slab and reinforcement meet the fire resistance required. Note the depth of slab assumed.

3. Design the ASBs using software. Choose fire engineered sections if fire protection is to be avoided. Ensure that the depth of slab covers the ASB by at least 30mm, or revise the slab depth (step 2) to be flush with the top of ASB, and provide reinforcing bars through the beam web.

4. Design any edge RHS beams using software. Design edge beams as non-composite to avoid the need to install U-bar transverse reinforcement. Design any Universal Beam edge members using resistance tables or software. Ensure the beam depth is compatible with the slab depth, or that it lies within a raised floor.

Typical section sizes

280 ASB 100 for 6 m span at 6 m centres

280 ASB 124 for 7.5 m span at 7.5 m centres

300 ASB 249 for 9 m span at 9 m centres.

Grade of steel ASBs are only available in S355 steel.

RHS *Slimflor* beams are available in S275 and S355.

Type of concrete Either normal weight concrete (NWC), 2350 kg/m^3 dry density, or lightweight concrete (LWC), 1850 kg/m^3 dry density, can be used.

NWC has better sound reduction, so is better for residential buildings, hospitals, etc.

LWC is better for overall building weight/foundation design, better span capability of slab, and has better fire insulating properties, enabling thinner slabs (10 mm less) to be used. It is not available in all parts of the UK. LWC is not considered suitable for directly-bonded floor coverings.

Overall floor zone Typically, 650 mm light services (with raised floor).

1000 – 1200 mm with air conditioning (and raised floor).

500 mm (min)

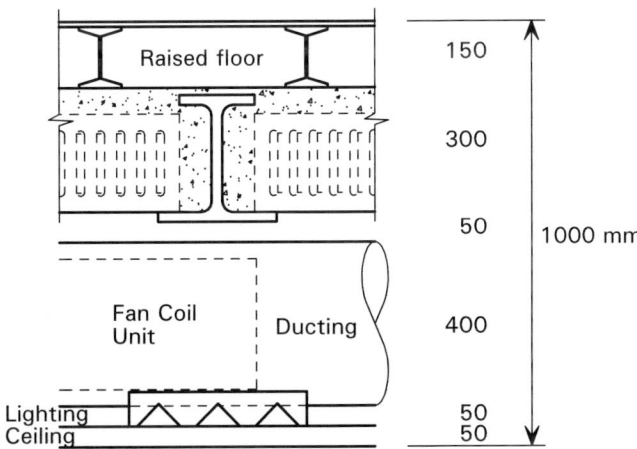

Figure 4.4 *Slimdek - Typical cross-section with air conditioning*

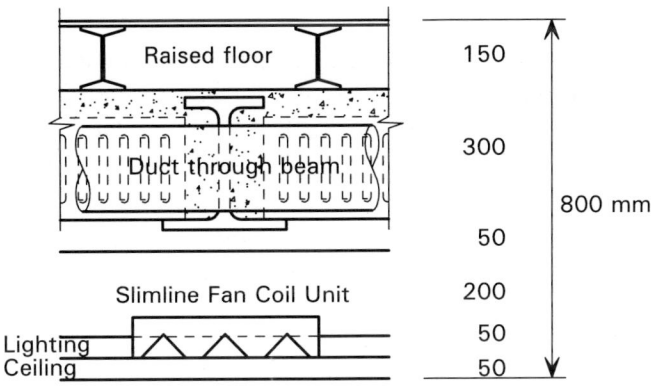

Figure 4.5 *Slimdek - Typical cross-section without air conditioning*

Fire protection Fire engineered ASBs with the web and top flange encased with concrete do not need fire protection for up to 60 minutes.

Thin web ASBs require fire protection for greater than 30 minutes - normally by board.

RHS *Slimflor* edge beams normally require fire protection for greater than 60 minutes. - normally by board.

Connections ASBs require end plate connections (typically, 6 or 8 bolt) to resist torsional loads. RHS *Slimflor* beams often use extended end plate connections to minimise the connection width.

Design guidance For details of the *Slimdek* system; *Slimdek* manual[23]

For design of ASBs; P175[24]

For the design of RHS *Slimflor* beams; P169[25]

Software Deep decking/slab: - Comdek, from www.corusconstruction.com

 - ASB: design software from www.corusconstruction.com

Edge beams: - RHS *Slimflor* software, from www.corusconstruction.com.

[Blank Page]

4.3 Cellular composite beams with composite slab and steel decking

Description Cellular beams are beams with openings at short regular intervals along their length. The beams are either fabricated from 3 plates or made from rolled sections. Openings, or 'cells', are normally circular, which are ideally suited to circular ducts, but can be elongated, rectangular or hexagonal. Cells may have to be filled in to create a solid web at positions of high shear, such as at supports or either side of point loads along the beam. The size and spacing of the openings can be restricted by the fabrication method, as well as by the required strength of the beams.

There are currently two companies who specialise in design and supplying these beams:

Westok Ltd, who supply cut and re-welded rolled sections, which can be of different weights or serial size, and can incorporate a camber when re-welded. These cellular beams generally have circular openings at regular centres.

Fabsec Limited, who supply beams fabricated from three plates with openings cut in the web. These beams can have a wide range of opening types and spacings, and can be supplied with a camber.

Cellular beams can be arranged as long-span secondary beams, supporting the floor slab directly, or as long-span primary beams supporting other cellular beams or conventional rolled section secondary beams.

Typical beam span range 10 – 18 m

Figure 4.6 *Long-span secondary cellular beams with regular circular openings*

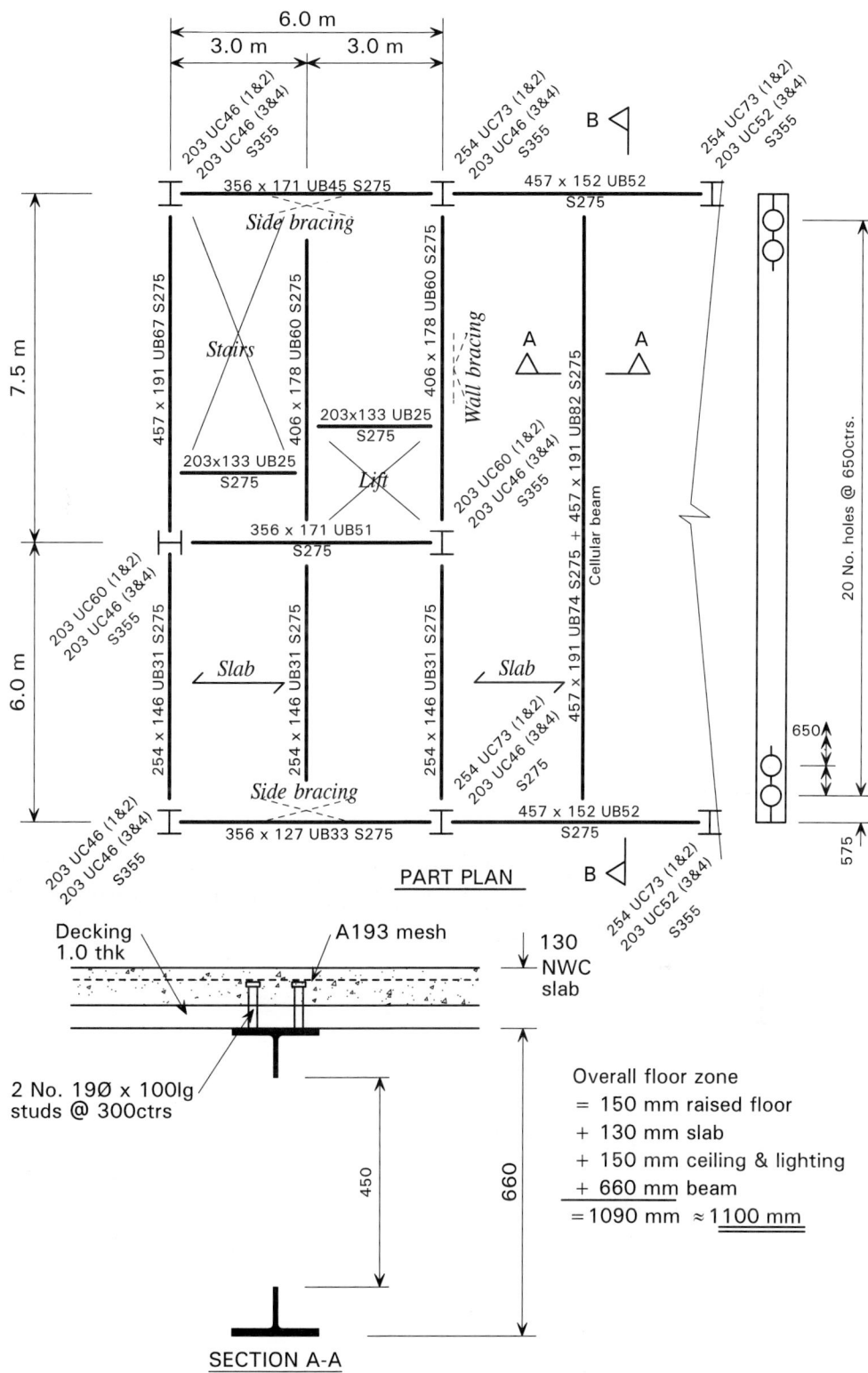

Figure 4.7 *Cellular beams (long-span secondary beams) and composite slabs – example of floor steelwork arrangement for a 4 storey rectangular plan building*

Main design considerations for the floor layout	Secondary beams should be spaced close enough to avoid propping the decking during construction. (3 – 4 m).
	Large (elongated or rectangular) openings should be located in areas of low shear, e.g. in the middle third of span for uniformly loaded beams.
	Consider integration of the services within the beam depth to minimise the overall floor zone.
Advantages	Long, column-free floor spans.
	Relatively lightweight beams compared with other long-span systems.
	Economic long-span solution.
	Precamber can be accommodated during the fabrication of the members.
Disadvantages	Increased fabrication costs compared with plain sections.
Services integration	Regular openings in the web allow ducts to pass through the beams. Larger items of equipment are located between the beams. Openings need to allow for any insulation around the services. Fabrication should be arranged to ensure web openings align through beams.
Governing design criteria for beams	Critical check may be elements within the beam – for example, the web posts between openings, particularly near high point loads or adjacent to elongated openings.
	The dynamic response of the floor may be critical in some circumstances.
	Opening size is ideally within 80% of the finished beam depth, and with a maximum opening length/depth ratio of 2.5. Stiffeners may be required for large openings.
Governing design criteria for decking/slab	Strength or deflection of the decking in the construction condition.
	Fire resistance (affects the concrete cover to the decking and mesh reinforcement size).
	Strength or deflection of the slab in the composite condition.
Design approach	1. Assume long-span secondary beams at 3 – 4 m spacing, supported by primary beams on a 6 m, 7.5 m or 9 m column grid
	2. Choose the decking and slab, using decking manufacturer's load tables or software. Assume LWC unless there is a directly-bonded floor covering. Assume C35 concrete, and unpropped decking during construction. Ensure the chosen slab and reinforcement meet the fire resistance required.
	3. Design the beams using manufacturer's software. Try shear studs at approximately 300 mm spacing on secondary beams (to suit trough spacing), and at 150 mm spacing on primary beams. Note that the orientation of the decking will differ between secondary and primary beams. As the services are likely to be integrated, ensure cell sizes and positions are agreed with the services engineer.
Typical section sizes	700 mm deep steelwork for 15 m span at 3.75 m centres.
	(Beam + slab depth) ≈ span/16-19.
Grade of steel	S275 and S355.

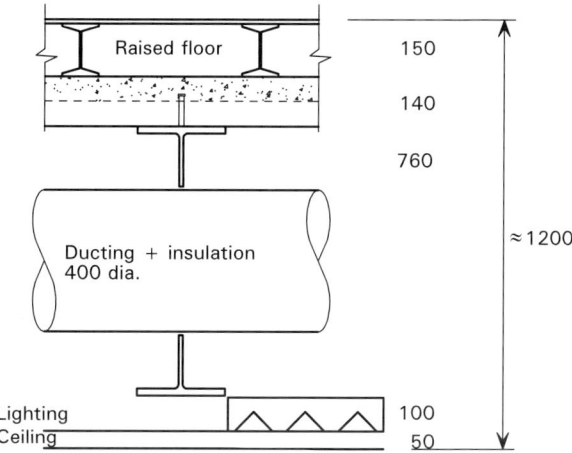

Figure 4.8 *Cellular beam - Typical cross-section*

Type of concrete	Either normal weight concrete (NWC), 2350 kg/m^3 dry density, or lightweight concrete (LWC), 1850 kg/m^3 dry density, can be used.
	NWC has better sound reduction, so is better for residential buildings, hospitals, etc.
	LWC is better for overall building weight/foundation design, better span capability of slab, and has better fire insulating properties, enabling thinner slabs (10 mm less) to be used. It is not available in all parts of the UK. LWC is not considered suitable for directly-bonded floor coverings.
Overall floor zone	1200 mm for 15 m span with 400 mm opening.
Fire protection	Intumescent paint, often off-site. References [26] and [27] give advice on fire protection of beams with web openings.
Connections	End plate, in shear only.
Design guidance	For choice of decking and composite slab design (including fire resistance); manufacturer's design tables.
	For best practice advice in design and construction; P300[20]
	For design charts and worked example for decking and beams; P055[21]
	For the basic design of orthodox cellular beams; P100[28]
Software	Cellbeam software from www.westok.steel-sci.org
	Fabsec software from www.fabsec.co.uk

Figure 4.9 *Fabricated beam with off-site fire protection*

4.4 *Slimflor* beams with precast concrete slabs

Description This is a slim floor system where the beams are contained within the structural floor depth. A steel plate (typically 15 mm thick) is welded to the underside of a UC section to make the *Slimflor* beam. This plate extends beyond the bottom flange by 100 mm either side, and supports the precast floor units. A structural concrete topping with reinforcement is recommended to tie the units together. The topping thickness should cover the units by at least 30 mm. If used without a topping, reinforcement should be provided through the web of the beam to tie the floor on each side of the beam together, to meet robustness requirements. LWC or NWC (10 mm aggregate) may be used, often C30 or C40.

A composite *Slimflor* beam can be achieved by welding shear connectors (normally 19 mm dia × 70 mm long) to the top flange of the UC. Reinforcement is then placed across the flange into slots prepared in the precast units, or on top of shallow precast units. If the steel beams are to be designed compositely, the topping should cover the shear connectors by at least 15 mm, and the precast units by 50 mm.

Edge beams are often designed as non-composite, with nominal shear studs provided to meet robustness requirements. These studs are usually site-welded through openings cast in the precast units. Composite edge beams require careful detailing of U-bar reinforcement into slots in the units, and a greater minimum flange width.

Only 152 UCs and 203 UCs are normally suitable as composite beams because the overall depth of the floor slab becomes impractical for larger serial sizes.

Precast units are precambered to cancel out dead load deflections between beams.

Typical beam span range 4.5 m to 7.5 m generally, although 10 m spans can be achieved.

Main design considerations for the floor layout Ideally, the span of the precast units and the beam span should be optimised to produce a floor thickness compatible with the *Slimflor* beam depth. Beams loaded on one side only are relatively heavy because of torsional loading. Torsional loading during construction will need to be checked. A central spine beam with precast units spanning to downstand edge beams will generally be more economic than parallel transverse *Slimflor* beams. RHS *Slimflor* edge beams may be used. Composite edge beams require careful detailing of U-bars around the shear connectors and into the precast units or structural topping – non-composite edge beams are usual.

Advantages Beams normally require no fire protection for up to 60 minutes fire protection.

Shallow floor zone – reduction in overall building height and cladding. Virtually flat soffit allows easy service installation and offers flexibility of internal wall positions.

Shear connectors can be welded off site, enabling larger stud diameters to be used and reducing site operations.

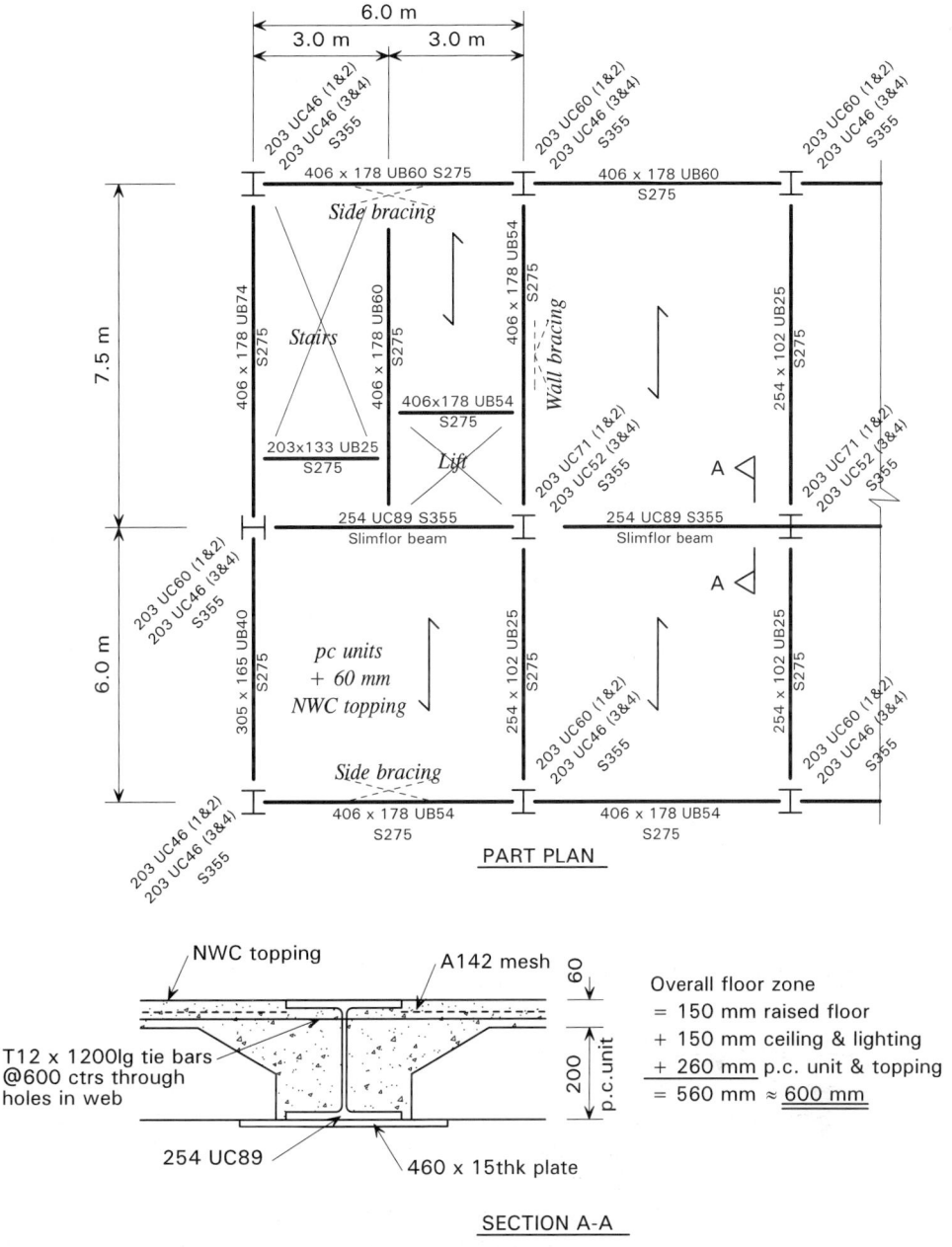

Figure 4.10 *Slimflor beams and precast concrete slabs with concrete topping flush with top flange*

Disadvantages	The steelwork is relatively heavy.
	Extra fabrication is involved in welding the plate to the UC. Connections require more detailing as the plate is wider than the column.
	Precast units involve more individual lifting operations than decking, which is delivered and erected in bundles. The erection sequence requires access for installation of the concrete units.
Services integration	Virtually flat soffit allows unrestricted access for services below the floor.
Governing design criteria for beams	Critical checks are usually the torsional resistance, combined torsion and lateral torsional buckling resistance (LTB) in the construction condition (when loaded on one side only), or LTB in the construction condition (with loads on both sides).
	Deflection may be critical with shallow beams.
Governing design criteria for precast units	Bending resistance.
	Shear resistance of hollow core units (for high applied shears, or for propped construction, consult plank manufacturer).
	Shape/dimensions of the end of the unit (rectangular or chamfered) to allow sufficient gap for free flow of concrete around the steel section (60 mm minimum between the concrete units and the steel is recommended).
	Detailing of transverse reinforcement around the beam shear studs and into the precast units, where composite action or improved fire resistance is required
	Length of the unit for end bearing (75 mm minimum for non-composite action and 60 mm minimum for composite action is recommended).
Design approach	1. Try 6 m, 7.5 m or 9 m grid.
	2. Choose precast concrete planks from manufacturer's data. Ensure these meet the required fire resistance. Longer spans are likely to be composite. Note the overall depth.
	3. Design the *Slimflor* beam using software. Beams may be non-composite or composite. Check that the cover to composite beams is at least 15 mm over the studs. If non-composite, allow for ties between the precast units through the web.
	4. Design the edge beams – either RHS *Slimflor* beams loaded on one side or downstand rolled sections. Design the edge beams as non-composite to avoid the need to install u-bar transverse reinforcement.
Typical section sizes	Beam ≈ 152 UC 37 + plate for 4.5 m span at 4.5 centres.
	Beam ≈ 203 UC 71 + plate for 6.0 m span at 6.0 centres.
	Beam ≈ 254 UC 167 + plate for 7.5 m span at 7.5 centres.
	Precast units ≈ 150 mm deep for 6 m span, 200 mm deep for 7.5 m span, 260-300 mm deep for 9 m span.
Grade of steel	S275 or S355.
Overall floor zone	600 mm with small services (with raised floor).
	1000 mm with air conditioning (with raised floor).

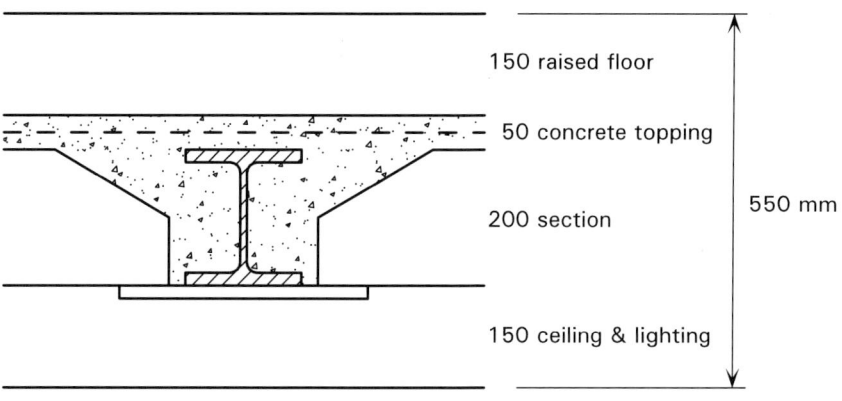

(a) Non-composite *Slimflor* beam with raised floor

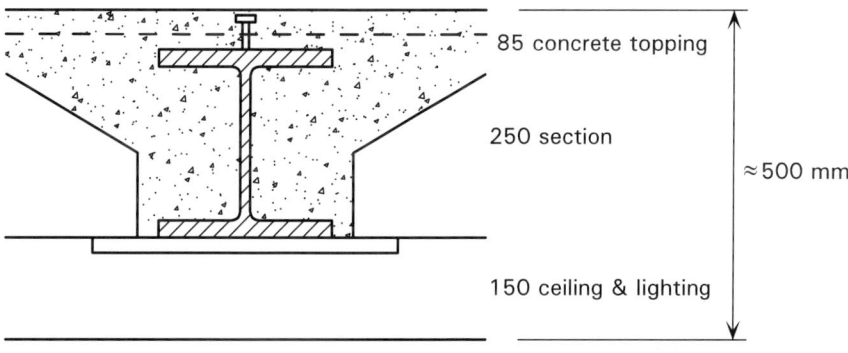

(b) Composite *Slimflor* beam without raised floor

Figure 4.11 *Slimflor construction – typical cross-sections*

Fire protection	The concrete encasement around the beam is normally sufficient for up to 60 minutes fire resistance without additional protection.
	For 90 minutes fire protection, an intumescent coating or board protection to the flange plate is required. Correct detailing of transverse reinforcement is required, particularly for hollow core units, where filling of the cores adjacent to the beam is necessary.
Connections	Full depth end plate connections are required to resist torsional loads in the construction condition.
Design guidance	For the design of slimflor beam; P110[29].
	For the design of composite *Slimflor* beams with precast slabs; P287[30].
	For the design of RHS *Slimflor* edge beams; P169[25].
	Precast units; manufacturers' design tables.
Software	*Slimflor* beams: *Slimflor* software, from www.corusconstruction.com.
	Edge beams: RHS *Slimflor* software, from www.corusconstruction.com.

4.5 Long-span composite beams and composite slabs with metal decking

Description This system consists of composite beams using rolled steel sections supporting a composite slab in a long-span arrangement of, typically, 10 to 15 m. Grids are either arranged with long-span secondary beams at 3 m to 4 m spacing supporting the slab, supported by short-span primary beams, or with short-span secondary beams (6-9 m span) supported by long-span primary beams. The depth of the long-span beams means that service openings, if required, are provided within the web of the beam. Openings can be circular, elongated or rectangular in shape, and can be up to 70% of the beam depth. They can have a length/depth ratio of up to 2.5. Web stiffeners may be required around holes.

Shear studs are normally positioned in pairs, with reinforcing bars placed transversely across the beams to act as longitudinal shear reinforcement.

Typical beam span range Long-span secondary beams: 9 m to 15 m span at 3 to 4 m spacing.

Long-span primary beams: 9 m to 15 m span at 6 to 9 m spacing.

Main design considerations for the floor layout Secondary beams should be placed close enough to avoid propping the decking (3 – 4 m).

Large (elongated or rectangular) openings should be located in areas of low shear, e.g. in middle third of the span for uniformly loaded beams.

Advantages Large column-free areas.

Disadvantages Deeper floor zones.

Heavier steelwork than some short-span solutions.

Fire protection required for 60 minutes fire resistance and above.

Services integration Service ducts pass through openings in the web of the beams

Larger plant can be situated between beams.

Governing design criteria for beams Critical checks are usually deflections and dynamic response. The combined response of primary and secondary beams should be checked. Shear resistance at openings, at supports or at point loads may be critical.

Governing design criteria for decking/slab Strength or deflection of the decking in the construction condition.

Fire resistance (affects the concrete cover to the decking and mesh reinforcement size).

Strength or deflection of the slab in the composite condition.

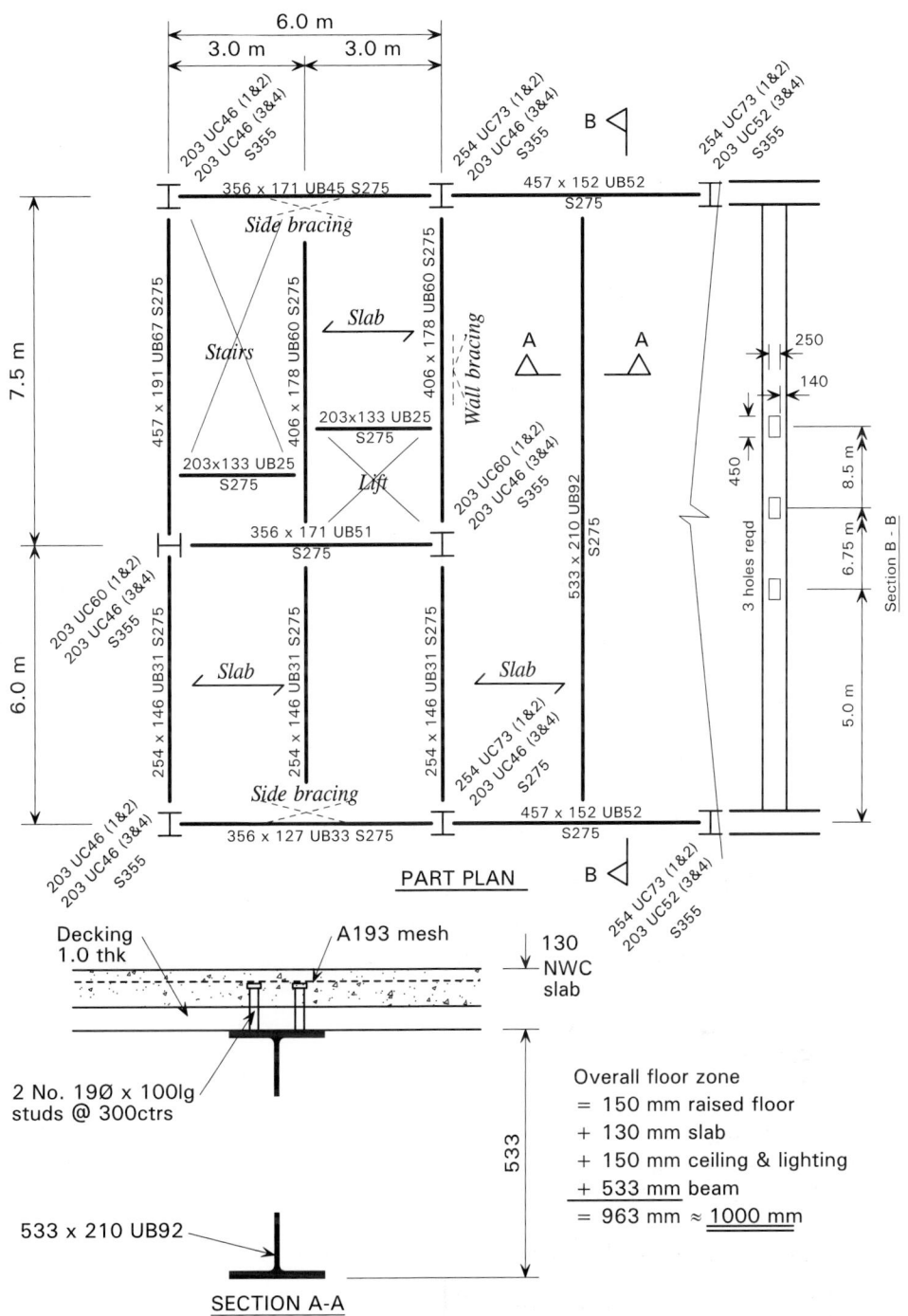

Figure 4.12 *Long-span composite beams (with web openings)*

Design approach	1. Try long-span secondary beams at 3 – 4 m spacing, on a 6 m, 7.5 m or 9 m column grid
	2. Choose the decking and slab, using decking manufacturer's load tables or software. Assume LWC unless there is a directly-bonded floor covering. Assume C35 concrete, and unpropped decking during construction. Ensure chosen slab and reinforcement meet the fire resistance required.
	3. Design beams using software. Try studs at approximately 300 mm spacing on secondary beams (to suit trough spacing), and at 150 mm spacing on primary beams. Note that the orientation of the decking will differ between secondary and primary beams. Ensure any holes in the web are of a size and location agreed with the services engineer, and allow for insulation around the services.
Typical section sizes	Composite section depth ≈ span/17-20.
	$533 \times 210 \times$ UB 92 for 13.5 m span at 3 m spacing.
Grade of steel	S275 or S355.
Type of concrete	Either normal weight concrete (NWC), 2350 kg/m^3 dry density, or lightweight concrete (LWC), 1850 kg/m^3 dry density, can be used.
	NWC has better sound reduction, so is better for residential buildings, hospitals, etc.
	LWC is better for overall building weight/foundation design, better span capability of slab, and has better fire insulating properties, enabling thinner slabs (10 mm less) to be used. It is not available in all parts of the UK. LWC is not considered suitable for directly-bonded floor coverings.
Overall floor zone	1000 mm for 13.5 m span (with 250 mm deep web opening)
	1200 mm for 15.0 m span (with 400 mm deep web opening)
Fire protection	Board, or intumescent coating (often applied off-site).
Connections	End plate connections, resisting shear only.
Design guidance	For choice of decking and composite slab design (including fire resistance); manufacturer's design tables.
	For best practice advice in composite design and construction; P300[20]
	For design charts and worked example for deck and beam; P055[21]
	For fire protection; the 'Yellow Book'[22]
	For advice on floor dynamics; P076[7].
Software	Slab design - Comdek software, available from www.corusconstruction.com
	- Deckspan software, available from www.rlsd.com
	- Multideck software, available from www. kingspanmetlcon.com/services/software/index.htm
	Beam design - BDES software, available www.corusconstruction.com

[Blank Page]

4.6 Composite beams with precast units

Description This system consists of rolled steel beams with shear studs welded to the top flange. The beams support precast concrete units with a structural concrete infill over the beam between the ends of the units, and often with an additional topping covering the units. The precast units are either hollow core, normally 150 - 260 mm deep, or they are solid planks of 75 mm to 100 mm depth.

At the supports, the deeper units are either chamfered on their upper face or notched down - to allow a thicker topping depth to fully encase the shear connectors. Narrow slits are created within the units during the manufacturing process to allow transverse reinforcement to be laid across the beams and be embedded in the precast units for approximately 600 mm either side (see Table 4.1 for recommended sizes). For hollow core units, the top of a number of discrete (not adjacent) cores need to be broken out during manufacture so that reinforcement can be placed and concreted into position.

The shear studs and transverse reinforcement allow the transfer of the longitudinal shear force from the steel section into the precast unit and the concrete topping, so that they can act together compositely. Composite design is not permitted unless the shear connectors are situated in an end gap (between the concrete units) of at least 50 mm. For on-site welding of shear connectors, a practical minimum end gap between concrete units is 65 mm. Stud capacity depends on the degree of confinement of the stud. LWC or NWC (10 mm aggregate) may be used in grades C30 or C40 for the topping. Hollow cores should be back-filled at the supports for a minimum length equal to the core diameter to provide a solid floor adjacent to the shear connectors, for effective composite action and adequate fire resistance.

Edge beams are often designed as non-composite, with nominal shear studs provided to meet robustness requirements. These studs are usually site-welded through openings cast in the precast units. Composite edge beams require careful detailing of U-bar reinforcement into slots in the units, and a greater minimum flange width.

Minimum flange widths are crucial for providing a safe bearing for the precast units and room for the shear studs – see below for minimum recommended values.

Temporary bracing providing lateral restraint is often required to reduce the effective length for lateral torsional buckling of the beam during the construction stage, when only one side is loaded. Full torsional restraint in the temporary condition may be difficult to achieve, unless deep restraint members with rigid connections are used, or by developing 'u-frame action' involving the beams, the restraint members and rigid connections.

Typical beam span range 6 - 9 m span beams, 6 - 9 m span precast units.

Main design considerations for the floor layout Maximise the span of the precast units.

A central spine beam with precast units spanning to edge beams will generally be more economic than precast units spanning between parallel transverse beams.

Beams that are parallel to the span of the precast units cannot be designed compositely.

Design edge beams as non-composite, but tied into the floor to meet robustness requirements.

Transverse reinforcement must be provided, as detailed in Table 4.1.

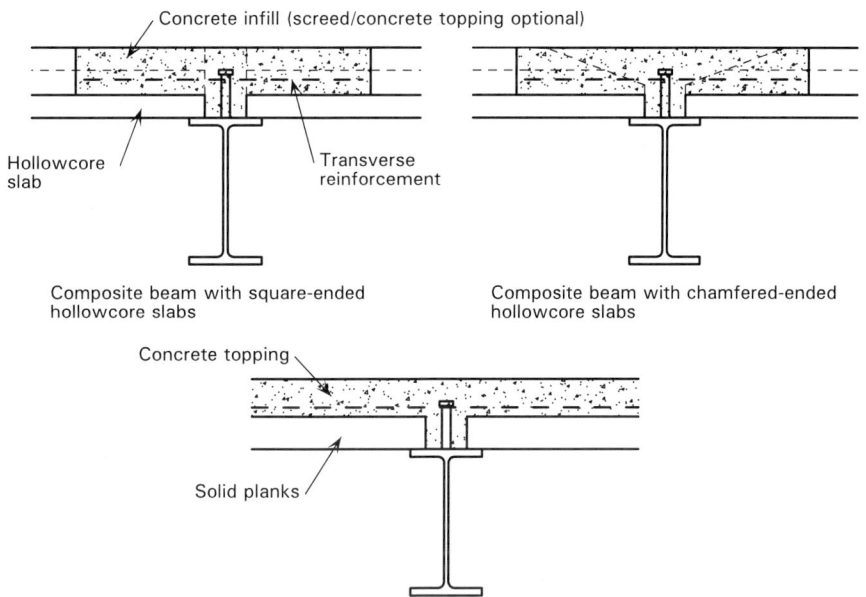

Figure 4.13 *Forms of composite beam with precast units*

Figure 4.14 *composite floor construction with precast concrete hollow core units, showing transverse reinforcement bars placed within open cores*

Table 4.1 *Recommended bar sizes for transverse reinforcement*

Slab depth	Bar sizes
Solid Planks	T10 @ 300 mm centres plus A142 mesh reinforcement
Hollow Core Units (up to 200 mm deep)	T16 @ 200 to 350 mm centres (unless full shear connection is provided, in which case T12 may be used)
Hollow Core Units (up to 260 mm deep)	T16 @ 200 to 350 mm centres

Advantages Fewer secondary beams, due to long-span precast units.

Shear connectors for most beams can be welded off site, enabling larger stud diameters to be used and fewer site operations. It is usually convenient to weld studs to edge beams on site.

Disadvantages The beams are subject to torsion and may need stabilising during the construction stage.

The precast units need careful detailing for adequate concrete encasement of shear connectors and installation of transverse reinforcement.

More individual lifting operations compared to the erection of decking, and the erection sequence requires access for installation of the concrete units.

Services integration Main service ducts are located below the beams with larger equipment located between beams.

Governing design criteria for beams Flange width for bearing and studs, stud size (site-welded or factory-welded)

The critical check is often torsional resistance and twist, or combined torsion and lateral torsional buckling resistance (LTB) in the construction condition (with loads on one side only).

Minimum flange width for bearing:

75 or 100 mm deep solid unit	Internal beam – 180 mm minimum
	Edge beam – 210 mm minimum
Hollow core unit	Internal beam – 180 mm minimum
	Edge beam – 210 mm minimum

Non- composite edge beam – 120 mm minimum

Governing design criteria for precast units Bending resistance.

Shear resistance of hollow core units.

Detailing of beam transverse reinforcement into units, where composite action or increased fire resistance is required.

Typical section sizes Beams:

Minimum rolled serial size is 406 × 178 UB for precast units with chamfered end and shop-welded connectors.

Minimum rolled serial size is 457 × 191 UB for square-ended precast units with shop-welded connectors.

Minimum rolled serial size is 533 × 210 UB for site-welded shear connectors.

Precast units: (approximate)

150 mm deep, 6 m span @ 2.5 kN/m^2

200 mm deep, 7.5 m span @ 3.0 kN/m^2

250 mm deep, 9 m span @ 5.0 kN/m^2

Design approach
1. Try 9 m grid.
2. Choose precast concrete planks from manufacturer's data. Ensure these meet the required fire resistance. Longer spans are likely to be composite. Note the overall depth.
3. Design the steel beam, using software or P287[30]
4. Design edge beams – as non-composite to avoid costly transverse reinforcement.

Grade of steel S275 or S355

Overall floor zone 900 mm including ceiling

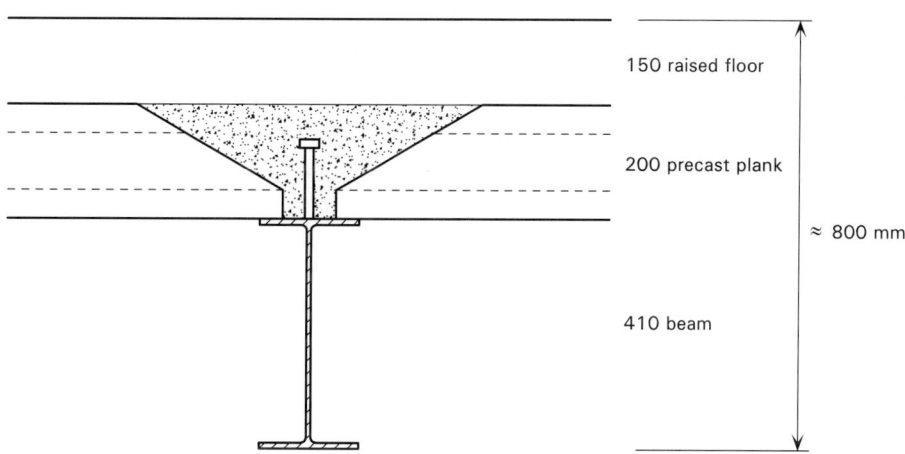

Figure 4.15 *Composite beam and precast concrete unit – typical cross-sections*

Fire protection Spray, board or intumescent coating to beam.

Transverse bars must be carefully detailed into the precast units – extending 600 mm into each unit. For 90 or 120 minutes fire resistance, a 50 mm (minimum) concrete topping is required.

Connections Full depth end plate connections (welded to the beam flanges) to cater for torsional loading.

Design guidance For beam design; P287

Precast units; manufacturer's design tables.

Software Software available from www.bison.co.uk

4.7 Non-composite beams with precast units

Description This system consists of rolled steel beams supporting precast concrete units. The precast units may be supported on the top flange of the steel beams, or, to reduce construction depth, supported on 'shelf' angles. Shelf angles are bolted or welded to the beam web, with an outstand leg long enough to provide adequate bearing of the precast unit and to aid positioning of the units during erection. Precast concrete units are generally grouted in position. The units may have a screed (which may be structural), or may have a raised floor. The precast units are either hollow core, normally 150-260 mm deep, or they are solid planks of 75 mm to 100 mm depth.

Temporary lateral bracing is often required to limit the effective length for lateral torsional buckling of the beam during the construction stage when only one side is loaded.

In order to meet robustness requirements, mesh and a structural topping may be required, or reinforcement concreted into hollow cores and passed through holes in the steel beam web. Tying may also be required between the concrete units and the edge beams.

Typical beam span range 6 m and 7.5 m grids are common for both beams and precast units.

Main design considerations for the floor layout Construction stage loading (planks on one side only) must be considered. Temporary bracing may be required.

Beams loaded on one side only in the permanent condition should either be avoided or designed for the applied torsion.

Central spine beams are common, with smaller edge beams. If edge beams carry significant cladding loads, or support inflexible cladding, deflection may be critical.

Advantages Fewer secondary beams, due to long-span precast units.

A simple solution involving basic member design.

Disadvantages The beams are subject to torsion and may need stabilising during the construction stage.

More individual lifting operations compared with the erection of decking, and the erection sequence requires access for installation of the concrete units.

Services integration Main service ducts are located below the beams with larger equipment located between beams.

Governing design criteria for beams Flange width or shelf angle width for bearing and erection access. A bearing of 75 mm is recommended (50 mm minimum). To allow for tolerances in the precast units and the erected steelwork, a gap of 30 mm between units is usually provided. When the top flange of a beam supports precast planks, the minimum flange width is therefore 178 mm.

Shelf angles should project at least 50 mm beyond the beam flange. When shelf angles are provided, 25 mm clearance is required between the end of the concrete unit and the beam flange, as shown in Figure 4.18.

The critical beam check is often torsional resistance and twist, or combined torsion and lateral torsional buckling resistance (LTB) in the construction condition (with loads on one side only).

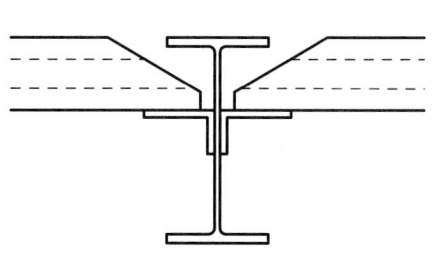 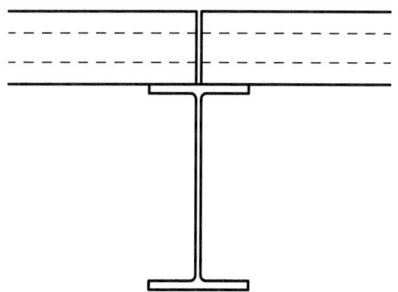

(a) Units sitting on shelf angles

(b) Units sitting on top of downstand beam

Figure 4.16 *Floor construction with precast concrete units in non-composite construction*

Figure 4.17 *Precast concrete units on steelwork*

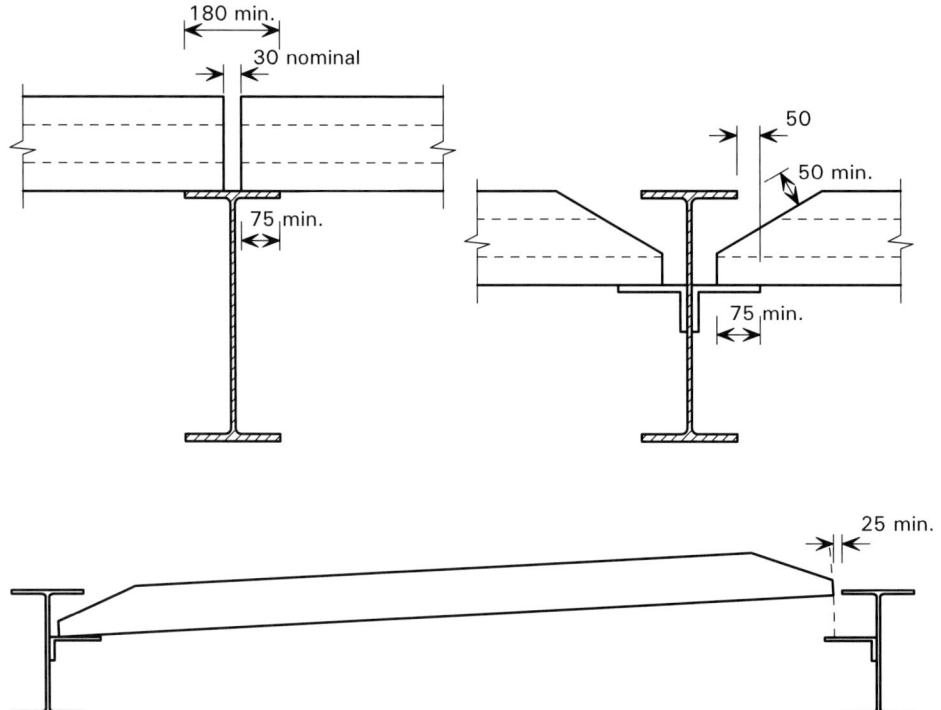

Figure 4.18 *Bearing and clearance requirements for precast units*

Governing design criteria for precast units
Bending resistance.
Shear resistance of hollow core units.

Typical section sizes
Beams:

When supporting precast planks on the top flange, the minimum rolled serial size is 406 × 178 UB.

Precast units: (approximate)

150 mm deep, 6 m span @ 2.5 kN/m^2

200 mm deep, 7.5 m span @ 3.0 kN/m^2

250 mm deep, 9 m span @ 5.0 kN/m^2

Design approach
1. Try 6 m or 7.5 m grid.
2. Choose precast concrete planks from manufacturer's data. Ensure these meet the required fire resistance. Longer spans are likely to be composite.
3. Design the steel beams, using software, or by simple manual calculation of the bending moment and deflection, with member resistances taken from the 'Blue book'[31]
4. Check the temporary construction condition, and consider temporary bracing as part of the erection method.

Grade of steel S275 or S355

Overall floor zone Approximately 800 mm including ceiling (7.5 m grid)

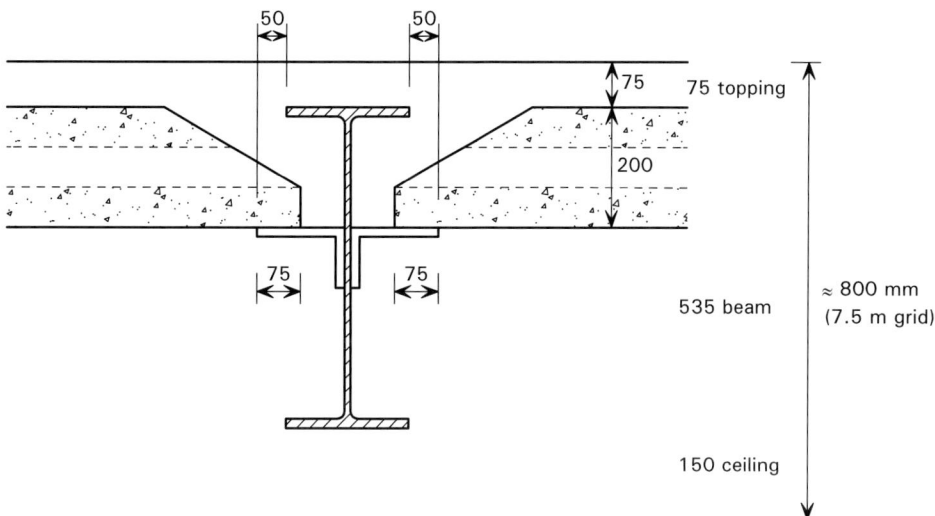

Figure 4.19 *Beam and precast concrete unit – typical cross-sections*

Fire protection Spray, board or intumescent coating to beam.

Connections Full depth end plate connections (welded to the beam flanges) are common, as the beams usually carry torsional loads in the construction condition.

Design guidance For beam design; P326[39]

For precast units; manufacturer's design tables.

Software Beam design – BDES software, available from Corus, can be used to design non-composite beams. BDES is available from www.corusconstruction.com

4.8 Beam connections

All the floor systems in Section 4 utilise simple connections – where the connections are assumed to behave as nominal pins, not developing significant moments. To realise this assumption in practice, the connection details must be flexible, to avoid moment transfer, and ductile, in order to accommodate the rotation that develops at the connection. In general, these connection characteristics are realised by detailing the connection with relatively thin connection components that are flexible and accommodate rotation. The industry standard connections are described in Section 4.8.1, and are 'partial depth' – the connection detail is approximately 60% of the beam depth.

Full depth connections are provided for floor members that are subject to torsion, such as asymmetric beams or *Slimflor* beams. For any floor solution, the possibility of torsional loading in the construction stage should be checked, as connections with torsional resistance, or temporary restraints may be required. In full depth connections, as described in Section 4.8.2, an end plate is welded to the beam flanges in addition to the web. Flexibility and ductility is maintained by using relatively thin end plates.

4.8.1 Simple connections

When connections are not subject to torsion, simple (vertical shear only) connections are usually detailed. In the UK, three standard beam connections are used, with the choice of detail left to the steelwork contractor. The standard connections are the flexible end plate, a fin plate or double angle cleats, shown in Figure 4.20

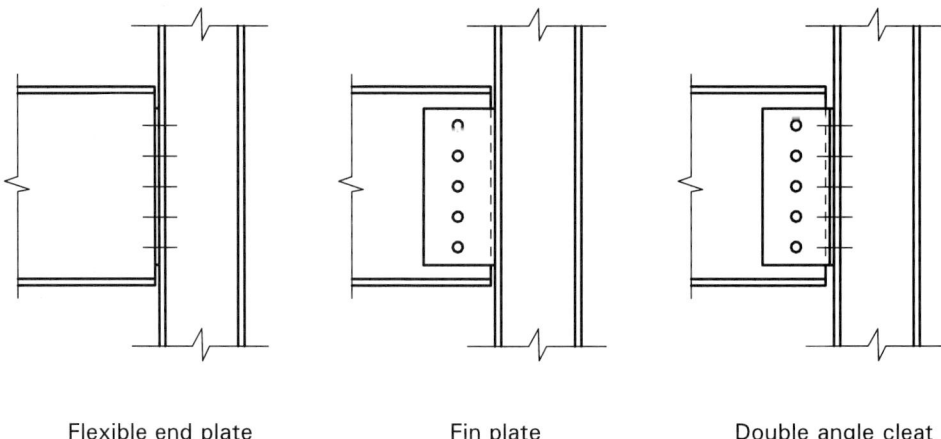

Flexible end plate   Fin plate   Double angle cleat

Figure 4.20 *Standard beam connections*

In general, flexible end plates are the most capable when considering vertical load, but less capable than other types of connection if considering substantial tying forces. Each connection type uses standard components. Connections to hollow sections are also straightforward, with the flexible end plate and double angle cleat connections using proprietary 'blind' fixings, or bolts using formed, threaded holes. Reference [41] gives full details.

Although Figure 4.20 shows connections to the flange of a column, the standard details can also be used for connections to column webs. Details of non-standard connections can be found in Reference [41].

Beam to beam connections also utilise the standard details, although the secondary beam will need to be notched, as shown by a flexible end plate example in Figure 4.21.

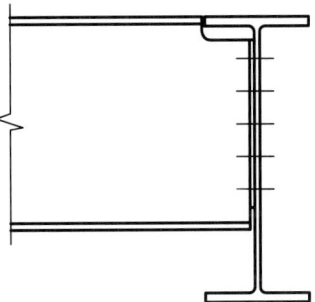

Figure 4.21 *Beam to beam connection*

A relatively new type of connection is the Quicon® connection, as shown in Figure 4.22. Special shouldered bolts are used with a T-piece fitting bolted to the support. The T-piece has keyhole shaped holes, allowing fast, safe erection. Further details are available from www.quicon.com.

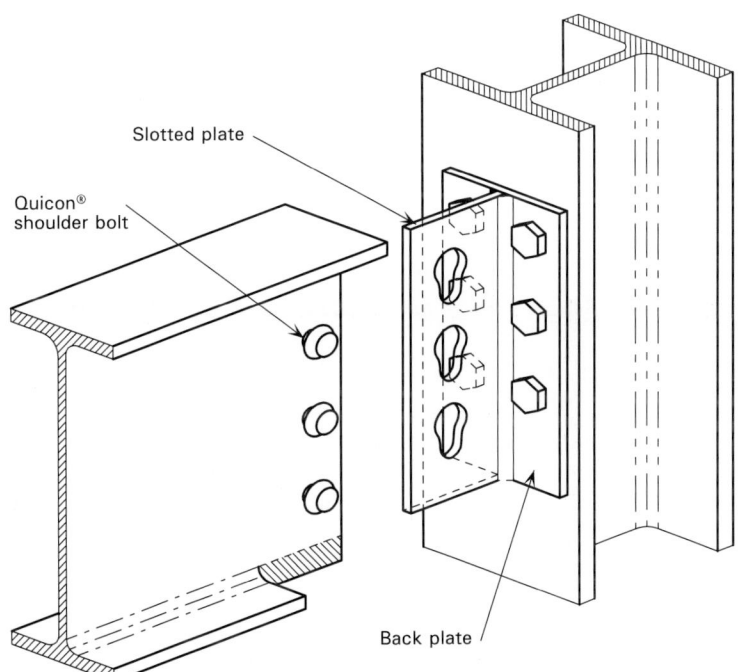

Figure 4.22 *Quicon® connection between beam and column*

4.8.2 Full depth end plates

When connections are subject to torsion, the connection is usually fabricated with a full depth end plate, as shown in Figure 4.23. In these connections, the end plate is welded around the full profile of the member.

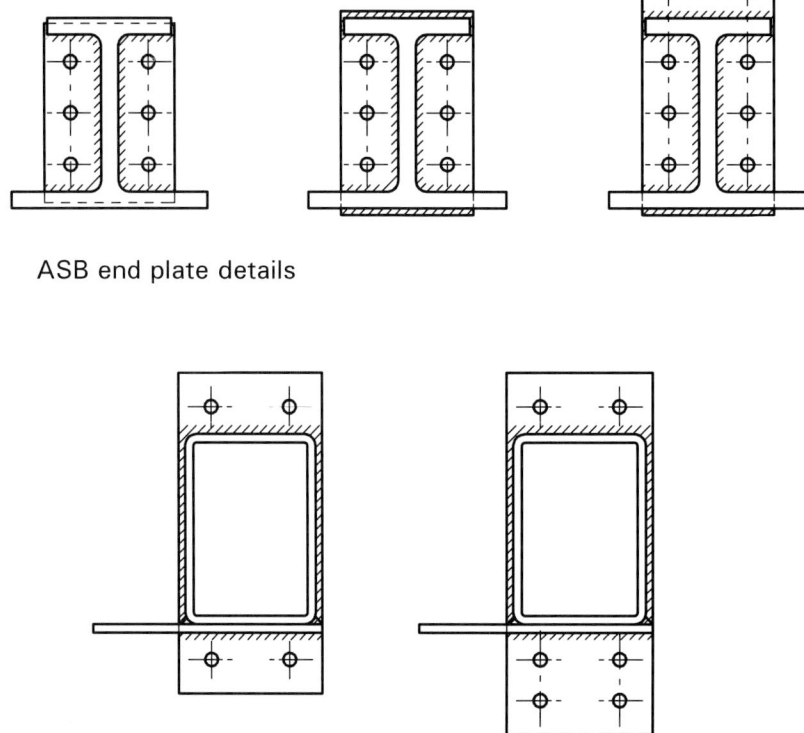

ASB end plate details

Figure 4.23 *Full depth end plates for ASB and RHS Slimflor beam*

It is usual practice for the steelwork contractor to design such connections. The structural designer should provide connection shears and torques for the relevant stages, i.e. during construction and in the final state. This is because for many members, torsion may be a feature at the construction stage, when loads are only applied to one side of the member. Other members, such as RHS *Slimflor* beams, will be subject to torsion at all stages.

For connections subject to torsion, the welds and the bolt group must be checked for the combined effects of the applied torsion and vertical shear.

When 'wide' members, such as ASBs, are connected to the minor axis of a column, a common detail is to weld a plate across the toes of the column flanges, as shown in Figure 4.24. In this situation, the welds between column flanges and the plate need to be designed for the combined effects of vertical shear and any torsion. Similar connection details are often used for RHS *Slimflor* edge beams, where the RHS is often offset from the centreline of the column to suit the detailing at the perimeter.

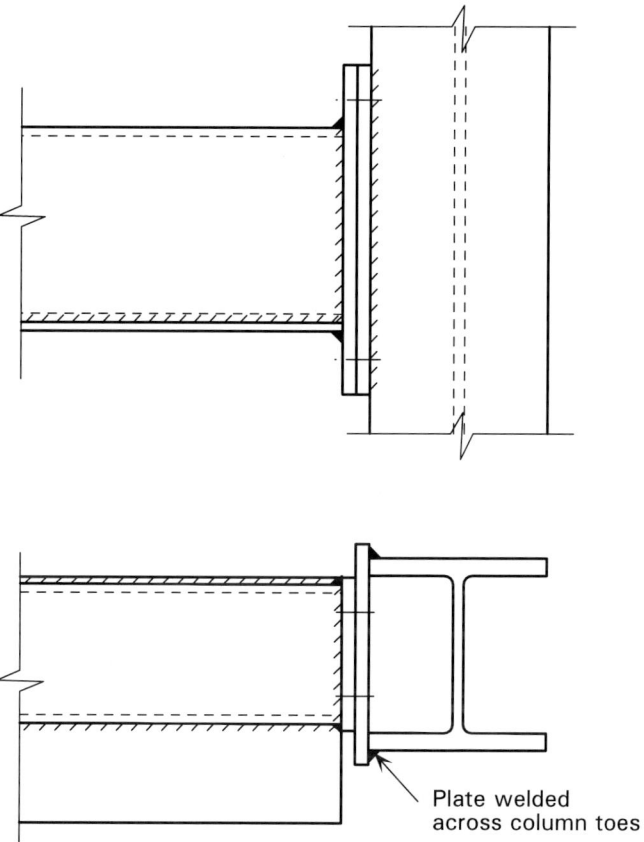

Figure 4.24 *Typical ASB connection to column minor axis*

5 COLUMNS

Columns in braced-frame multi-storey buildings are usually hot rolled Universal Column (UC) sections. UC sections provide good compression resistance and connections of beams to the webs and flanges of UC sections is straight forward.

Rectangular or circular structural hollow sections are sometimes used, often for aesthetic reasons. Hollow sections can be concrete-filled to achieve higher axial compression resistances and also to improve fire resistance.

The column size required will depend on the floor system used, the column spacing and the number of floors. Typical column sizes for short-span composite beams with composite slabs are given in Table 5.1 below. The size of column required is obviously dependent on the number of storeys that it is required to support.

Table 5.1 *Typical column sizes (short-span composite beams)*

Number of floors supported by column section	Universal Column (UC) serial size
1	152
2 – 4	203
3 – 8	254
5 – 12	305
10 – 40	356

For ease of construction, columns are usually erected in two, or sometimes three, storey sections (i.e. approximately 8 to 12 m lengths). Column sections are joined with splices, as described in section 5.4.

5.1 Column loading

Columns are subjected to a combination of axial loads and modest moments from the floor beams, as described in Section 5.2. To calculate the design forces and moments applied to a column, all beams supported by the column should be considered fully loaded (i.e. 1.4 × dead load + 1.6 × imposed load). The application of the Notional Horizontal Forces (NHF), as discussed in Section 7, will increase the axial loads in the columns that form part of the bracing system. For initial design purposes, the additional load due to the NHF can be ignored.

BS 6399-1[16] gives a reduction factor that may be applied to the total imposed floor load carried by a column, where the column carries more than one floor. The reduction is related to the number of floors supported by the column section being considered. Table 2 from BS 6399-1, which gives the reduction values, is reproduced as Table 5.2 below.

Table 5.2 *Reduction in total distributed imposed floor loads (from BS 6399-1)*

Number of floors with qualifying loads for reduction carried by member under consideration	Reduction in total distributed imposed load on all floors carried by the member under consideration
1	0 %
2	10 %
3	20 %
4	30 %
5 – 10	40 %
Over 10	50 % (maximum)

Not all imposed floor loads qualify for the reduction described above. Imposed floor loads that do not qualify for the reduction are:

- Loads that have been specifically determined from knowledge of the proposed use of the structure. This would be the case if loads other than the general, uniform floor loadings found in BS 6399-1 or equivalent, (see Section 3.1.2) have been used.

- Loads due to plant or machinery.

- Loads due to storage.

If the floor loads are based on a general requirement such as "4 plus 1" or "5 plus 1", the reductions may be applied.

5.2 Column design

Braced frame multi-storey construction comes within the scope of "simple" construction, as defined in BS 5950-1. For columns in simple structures, pattern floor loading need not be considered. The design of columns in simple construction should be carried out to Clause 4.7.7 of BS 5950-1. This clause contains simplified design rules for columns that take account of:

- Nominal moments applied by the eccentricity of loads at the connections (see Figure 5.1).

- The relative stiffnesses of column sections at a splice.

- The shape of the column bending moment diagram.

- The combined effects of axial load and moment.

Figure 5.1 shows the nominal moments applied to columns in simple construction. The Code assumes that 'nominal moments' are introduced by the beam end reactions acting at an assumed eccentricity. This relieves the designer of the necessity of trying to calculate the actual moments in the column. Beam end reactions should be taken as acting at 100 mm from the face of the steel column. Moments are not introduced into the column when the column is symmetrically loaded, and the column is therefore designed for axial load alone. Often, only columns on the edge of the structure will have out-of-balance loading. Most columns within a structure will be symmetrically loaded, and designed for axial load only.

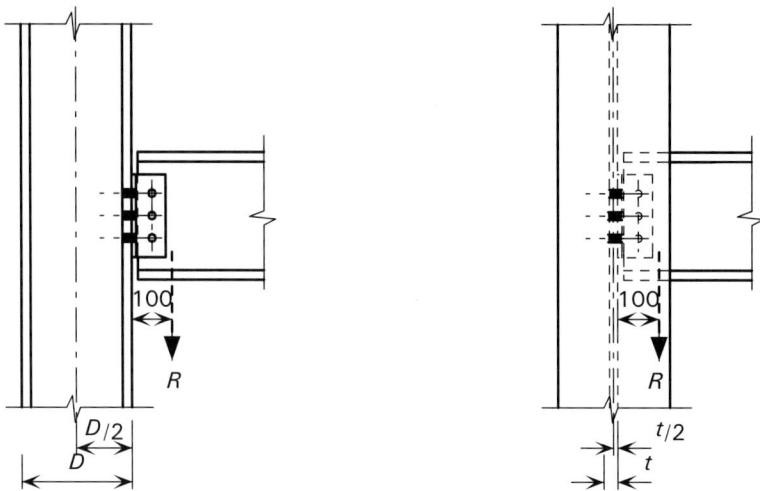

Nominal moment = $R(D/2 + 100\text{ mm})$ Nominal moment = $R(t/2 + 100\text{ mm})$

Figure 5.1 *Nominal moments from floor beams*

The distribution of nominal moments to the upper and lower column sections is carried out in proportion to their stiffness, except where the ratio of stiffnesses (I/L) does not exceed 1.5, when the moments may be shared equally. Figure 5.2 illustrates the assumed distribution of moments.

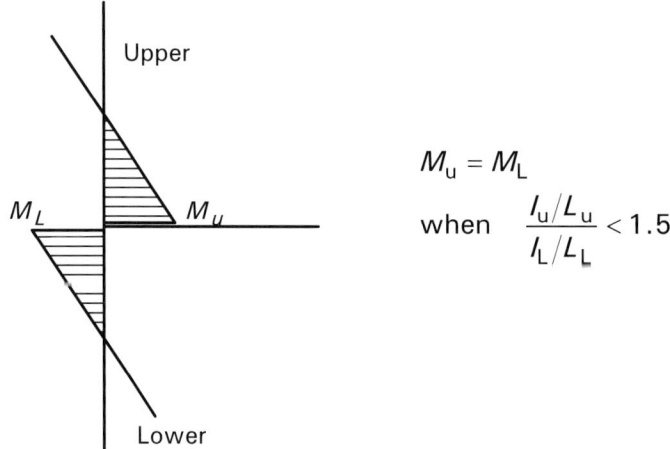

Figure 5.2 *Distribution of nominal moments from floor beams*

To take account of the combined effects of axial loads and moments the following single interaction expression is used:

$$\frac{F_c}{P_c} + \frac{M_x}{M_{bs}} + \frac{M_y}{p_y Z_y} \leq 1$$

where:

F_c is the compressive force due to axial load

P_c is the compression resistance

M_x is the nominal moment about the major axis

M_{bs} is the buckling resistance moment for simple columns

M_y is the nominal moment about the minor axis

p_y is the design strength

Z_y is the elastic modulus about the minor axis.

In calculating the buckling resistance moment M_{bs} for columns in simple construction, BS 5950-1 allows the value of the equivalent slenderness λ_{LT} to be calculated using a simplified expression, as given below, rather than the more complex formula used for unrestrained beams.

$$\lambda_{LT} = 0.5 \, L \, / \, r_y$$

where:

L is the distance between levels at which the column is laterally restrained in both directions

r_y is the radius of gyration about the minor axis.

A complete explanation of how Clause 4.7.7 should be applied is included in *Introduction to steelwork design to BS 5950-1:2000*[32].

5.3 Initial sizing

The following steps can be used for initial sizing of columns in simple construction.

Step 1. Determine the axial load in column F.

F = (1.4 × dead load + 1.6 × imposed load) × floor area supported by column × number of floors supported by column.

Step 2. If the column is an edge column, or has asymmetric loading, increase the axial load to allow for the effect of nominal moments. The more floors the column supports, the smaller the significance of the nominal floor moments. For a column supporting 2 floors multiply F by 1.25, for 4 floors multiply F by 1.15, for 6 floors or above multiply F by 1.05.

Step 3. Using an effective length equal to the storey height, select a column section with a minor axis compression resistance (P_{cy}) of at least the value calculated in Step 2. Tabulated values of axial resistance for column sections are given in the "Blue Book"[31].

5.4 Splices

Column splices in braced multi-storey construction are usually provided every two or three storeys. This results in convenient lengths for fabrication, transport and erection, and gives easy access from the adjacent floor for bolting up on site. The provision of splices at each storey level is seldom economical since the saving in column material is generally far outweighed by the material, fabrication and erection costs of making the splice.

5.4.1 Splice position

Traditionally, column splices were located about 600 mm above the floor level so that they were at a convenient height for fixing of bolts during erection, and where the internal moments were small. More recent practice is to locate the splice so that edge protection and handrails can be attached at 1100 mm above the floor level, as shown in Figure 5.3.

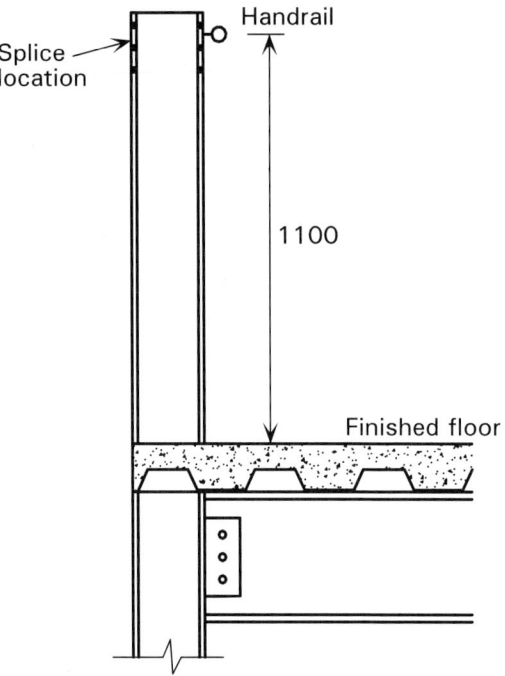

Figure 5.3 *Splice located to support perimeter scaffold*

At some splices within a multi-storey frame, it is common for the upper column to be a lighter section than the lower column section. It is good practice to ensure that this change of column section is not greater than one serial size. Standard details for column splices are presented in SCI publication, *Joints in steel construction: Simple connections* (P212)[41].

Ideally, a splice in a compression member should be positioned close to a restraint, or, if the member is continuous, at a point of inflexion of the buckled shape. (A splice within 600 mm of a restraint is considered "close" to a restraint.) Where splices are located elsewhere, special consideration needs to be given to the design of the splice to allow for the internal moments.

5.5 Splice design

The basic requirements are that the splice must be both:

- stiff enough to avoid reducing the buckling resistance of the member below that required, and also

- strong enough to transmit the forces and moments in the member.

Providing a splice that is more flexible than the member itself is considered to be bad practice, because it is potentially unsafe. It is a complex matter to calculate the buckling resistance of a member in which the splice does not have at least the same stiffness as the member. The design of splices is covered in detail by Advisory Desk Note 243[33].

SCI publications, P212 contains series of standard splice details with capacities.

For Universal Column sections, there are two forms of splice; bearing and non-bearing, both of which are shown in Figure 5.4

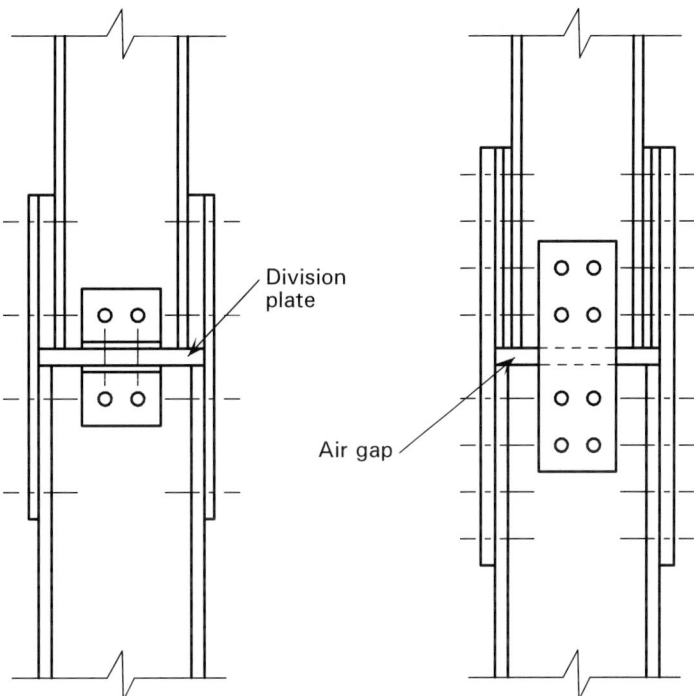

Figure 5.4 *Splices in Universal Column Sections*

In the bearing type of splice, the axial load is transferred in bearing between the column lengths either directly (same serial size) or via a division plate. The bolts and plates provide continuity of stiffness across the splice, and may be designed to resist a tension force if the rules covering frame robustness are applicable.

In the second type of splice, all the load is transferred between column lengths by the cover plates, which will generally result in a much larger and more expensive detail. For this reason, the bearing type splice is recommended. The ends of the columns do not need machining – a saw-cut end is satisfactory, and a small gap between bearing surfaces is in fact permitted in the *National Structural Steelwork Specification* (NSSS) [34]. The division plate is sized by assuming a spread of load at 45°, as shown in Figure 5.5

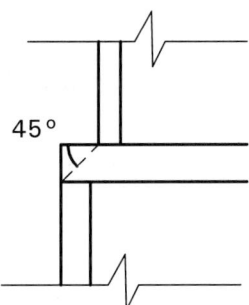

Figure 5.5 *Bearing type splice – division plate*

With external cover plates as shown in Figure 5.4, splices can be bulky, and inconvenient for architectural reasons. To reduce the overall size, splices are often detailed either with countersunk holes in the splice plates, or using internal cover plates and countersunk holes in the column section, as shown in Figure 5.6. In this latter detail, the splice can be detailed entirely within the profile of the column section.

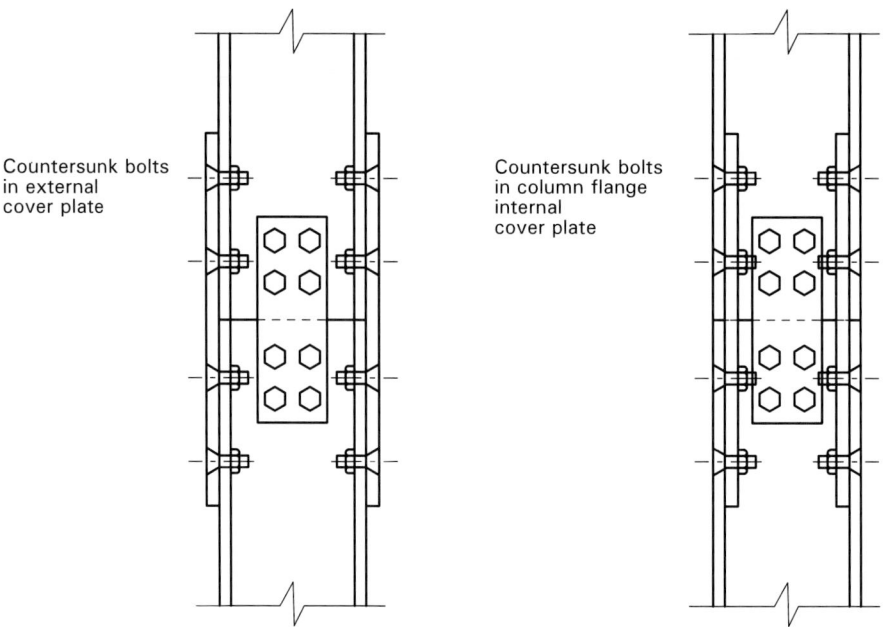

Figure 5.6 *Compact splice details (both are bearing splices)*

Designers should note that if different serial size columns are to be spliced, packs will invariably be required, as shown in Figure 5.7. Depending on the pack thickness and numbers of packs, the shear capacity of the bolts may be reduced; designers are referred to Clauses 6.3.2.2 and 6.3.2.3 of BS 5950-1.

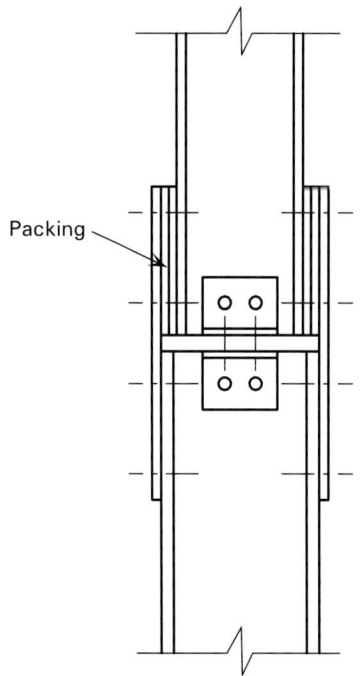

Figure 5.7 *Splice connection with packing*

For hollow sections, splices are generally achieved with a cap and baseplate detail as shown in Figure 5.8.

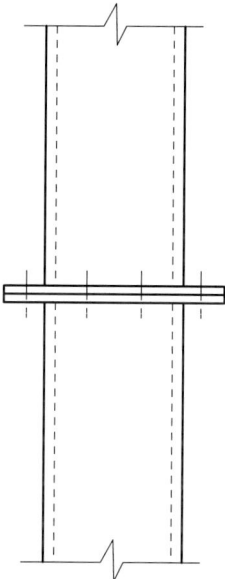

Figure 5.8 *Splice in hollow section column*

5.6 Column bases

Base details invariably comprise a square or rectangular plate, welded to the column section. It is recommended that four holding down bolts be provided, outside the section profile. This arrangement of bolts is preferred for three reasons:

- the wider spacing of the bolts increases stability in the temporary condition
- the bolt arrangement makes plumbing the column simpler
- columns are usually located on a central pack of shims, which can impose a significant point load. If holding down bolts are closely spaced, and in conical sleeves, there may be little remaining concrete. Wider spacing of the bolts leaves an increased block of foundation concrete between the bolts to support the point load.

Despite the four bolt arrangement, details such as those in Figure 5.9 are still considered as nominal pins when designing the frame.

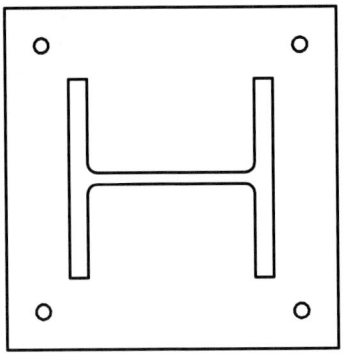

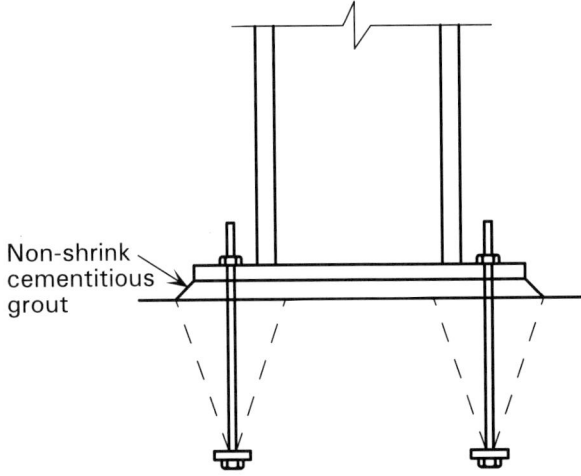

Figure 5.9 *Typical base detail for Universal Column*

Full details of standard bases to Universal Column sections and to circular, square and rectangular sections can be found in *Joints in steel construction: Simple connections*[41].

Holding down bolts

Holding down bolts are usually placed in conical sleeves, as shown in Figure 5.9. An anchor plate is generally provided for each bolt, although a common plate or angle section may connect two or more bolts. Holding down bolts should be moved as the concrete cures, to allow lateral movement when the steel is located. If this is not done, the bolts will be held so rigidly that they cannot be inclined to adjust the alignment of the steel base. Bolt holes through the base plate are made 6 mm oversize to accommodate the potential for the holding bolts to be inclined. Best practice is to have all holding down bolts of the same grade, because once the concrete has been cast, errors can be expensive to rectify.

5.7 Bases to braced bays

The base detail (and foundation) at the foot of a braced bay is not straightforward, as there will be a high horizontal shear and the possibility of uplift. Depending on the bracing arrangements, these effects may be on different bases, or both applied to the same base in combination.

Ideally, the steel base detail, the foundation and its reinforcement should all be developed together. The following guidance is merely one approach – other solutions are possible.

Horizontal shear

In combination with axial compression, designers often consider that horizontal shear is carried by friction between the steel and the foundation. A coefficient of friction of 0.3 is assumed; meaning the resistance to horizontal shear is 30% of the vertical compression. This approach is obviously inappropriate in combination with uplift.

Carrying horizontal shear via the holding down bolts is not recommended, as the bolts will experience bending, and thus have a reduced capacity. Designers are referred to Clause 6.3.2.2 of BS 5950-1 as an indicator of the reduction in capacity.

When significant horizontal shear must be transferred to the foundations, the column bases are either set in reinforced pockets in the base (see Figure 5.10) which are subsequently concreted, or a stub member can be welded on the underside of the base plate, which locates in a pocket in the foundation, as shown in Figure 5.11

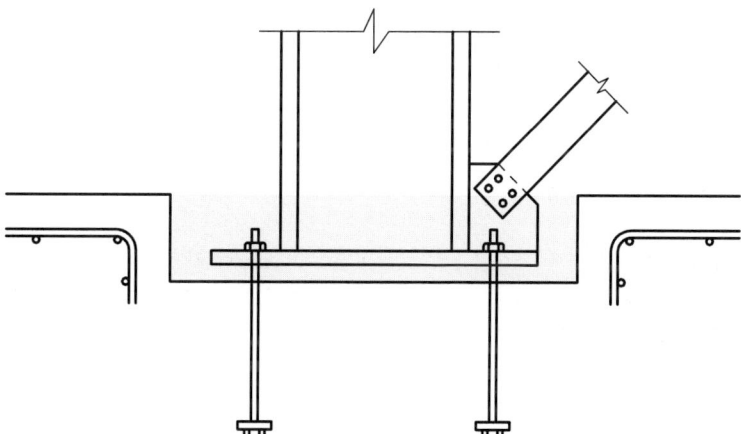

Figure 5.10 *Base located within reinforced pocket*

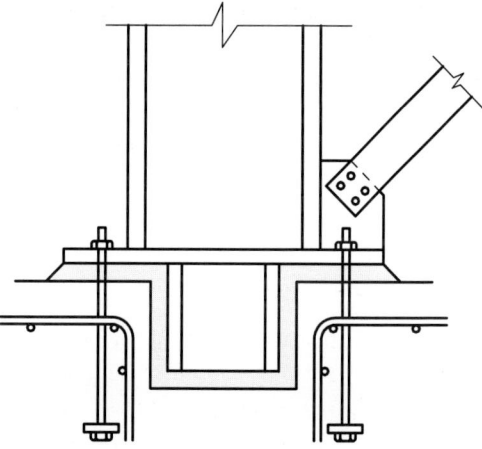

Figure 5.11 *Base with stub member in pocket*

The horizontal shear capacity of the stub is taken as the projected area multiplied by $1.5f_{cu}$, where f_{cu} is the characteristic strength of the concrete. The factor of 1.5 accounts for the high degree of constraints of the concrete (or grout), and accounts for transfer of force via both flanges of a universal section[35].

The stub (and weld) is usually designed for horizontal shear alone, assuming no bending occurs.

Uplift

Ideally, holding down bolts in uplift should be designed at the same time as the reinforced foundation, enabling a properly reinforced solution with the bolts an integral part. *Joints in steel construction: Moment connection*[36] has a comprehensive approach to calculating the pull-out capacity of a bolt and washer plate assembly. More traditional approaches, based on the pull-out capacity of a cone of concrete, assuming minimum reinforcement in the base, can be found in *Holding down systems for steel structures*[37] and *Steel designer's manual*[38].

6 BRACING SYSTEMS

6.1 Introduction

Two systems of bracing must be considered in multi-storey design. Vertical bracing resists the lateral loads and provides overall stability to the structure. Loads must be carried to the vertical bracing by ensuring that the floors and roof act as horizontal diaphragms. Usually, the floor system will be sufficient to act as a diaphragm without the need for additional steel bracing. At roof level, bracing, often known as a wind girder, may be required to carry the horizontal loads at the top of the columns, if there is no slab at roof level.

6.2 Horizontal diaphragms

As vertical bracing is generally provided in discreet locations (typically on the external elevations as shown in Figure 6.1), the floors and roof plates are required to act as a diaphragm to carry loads from intermediate frames to the braced frames.

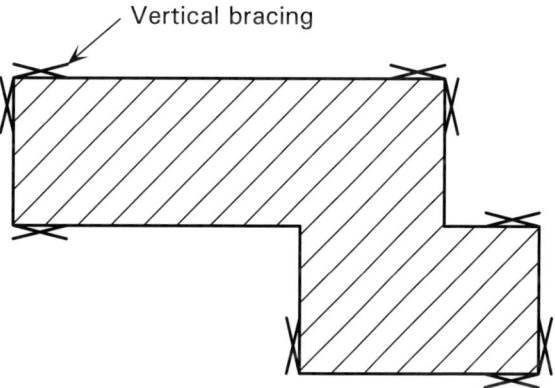

Figure 6.1 *Typical arrangement of vertical bracing*

Without such diaphragms, the intermediate frames could deflect laterally relative to the braced frames, as shown in Figure 6.2.

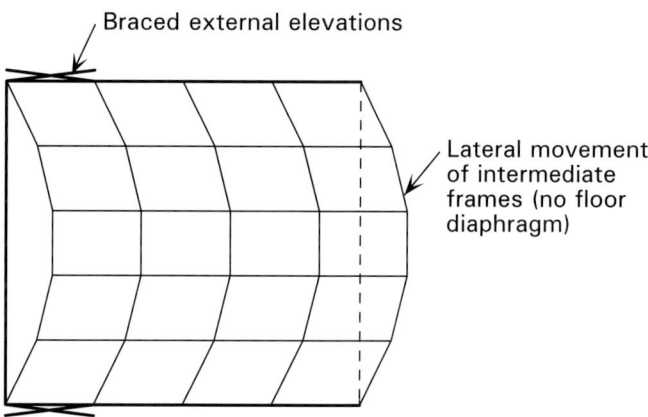

Figure 6.2 *Potential behaviour of frames without floor diaphragms*

All floor solutions involving permanent formwork such as metal decking fixed by through-deck welding to the beams, with in-situ concrete infill, provide an excellent rigid diaphragm to carry horizontal loads to the bracing system.

Floor systems involving precast concrete planks require proper consideration to ensure adequate load transfer. The coefficient of friction between planks and steelwork may be as low as 0.1, and even lower if the steel is painted. This will allow the slabs to move relative to each other, and to slide over the steelwork. Grouting between the slabs will only partially overcome this problem, and for large shears, a more positive tying system will be required between the slabs and from the slabs to the steelwork.

Connection between planks may be achieved by reinforcement in the topping. This may be mesh, or ties may be placed along both ends of a set of planks to ensure the whole panel acts as one. Typically, a 10 mm bar at half depth of the topping will be satisfactory.

Connection to the steelwork may be achieved by two methods:

- Enclose the slabs by a steel frame (on shelf angles, or specially provided constraint) and the gap filled with concrete.

- Provide ties between the topping and an in-situ topping to the steelwork (known as an 'edge strip'). The steel beam has some form of shear connectors to transfer load between the in-situ edge strip and the steelwork.

If plan diaphragm loads are transferred to the steelwork via direct bearing (typically the slab may bear on the face of a column), the capacity of the connection should be checked. The capacity is generally limited by local crushing of the plank. In every case, the gap between the plank and the steel should be made good with in-situ concrete.

Timber floors and floors constructed from precast concreted inverted tee beams and infill blocks (often known as 'beam and pot' floors) are not considered to provide an adequate diaphragm. In both cases, a system of horizontal steel bracing is recommended, or some other means to ensure that frames are properly braced to each other. A horizontal bracing system may need to be provided in both directions.

6.3 Horizontal bracing design

Typically, horizontal bracing systems span between the 'supports', which are the locations of the vertical bracing. This arrangement often leads to a truss spanning the full width of the building, with a depth equal to the bay centres, as shown in Figure 6.3.

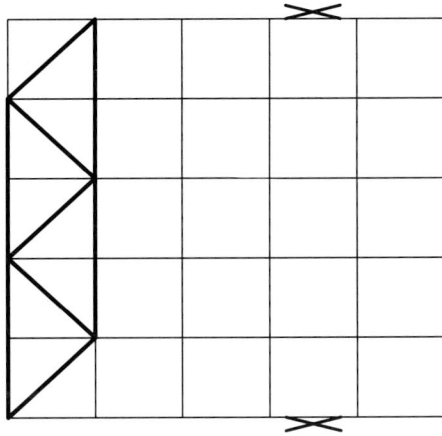

Figure 6.3 *Typical floor bracing arrangement*

The floor bracing is frequently arranged as a warren truss, or as a pratt truss, or with crossed members. In all cases, the lateral load applied to the truss is usually taken as the lateral load (pressure, drag and suction of wind load) applied from half a storey above and half a storey below the floor level being considered, as shown for both floors and roof in Figure 6.4.

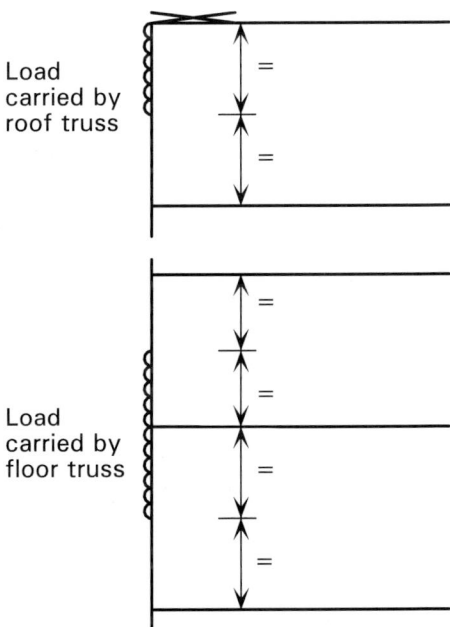

Figure 6.4 *Nominal allocation of load to floor trusses*

For simplicity, the reaction at the end of the horizontal truss can be calculated, and this resolved reaction used to design the critical bracing member, as shown in Figure 6.5. For simplicity, the same member size may be used throughout the bracing system.

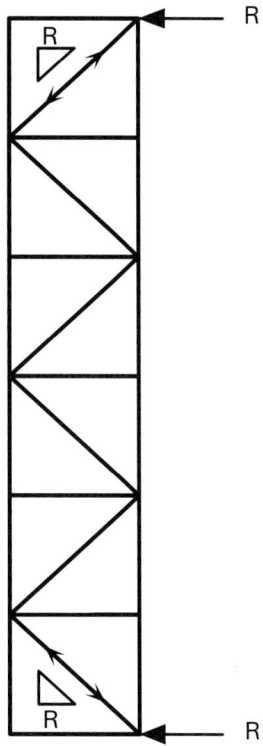

Figure 6.5 *Simple approach to bracing design*

6.4 Vertical bracing

Generally, at least two sets of bracing are provided in each orthogonal direction. It is preferable to locate bracing at or near the extremities of the structure, in order to resist any torsional effects (twist on plan) from wind or lateral loads. Bracing systems concentrated at one location may be subject to high torsional loads, particularly if the concentration of bracing is located away from the centroid of the building. Where the sets of bracing are identical or similar, it is sufficient to assume that external loads (typically wind) and stability effects (see Section 7) are split equally amongst the bracing systems in the orthogonal direction under consideration. Although bracing is often located on the external elevations, this may not always be possible. Bracing may be provided around service cores, or lift shafts, or stair wells, and bracing members may be concealed within cavity walls. Occasionally, architectural arrangements may have to be modified to facilitate the location and concealing of bracing members. In some circumstances, additional columns may have to be introduced solely to act as part of the bracing system.

Where the stiffnesses of the bracing systems differ (different truss 'depths' for example), or the bracing systems are located asymmetrically in the horizontal direction, as shown in Figure 6.6, an equal distribution of load should not be assumed. The loads carried by each bracing system can be calculated by assuming the floor is a stiff beam and the bracing systems are spring supports, as shown in Figure 6.6.

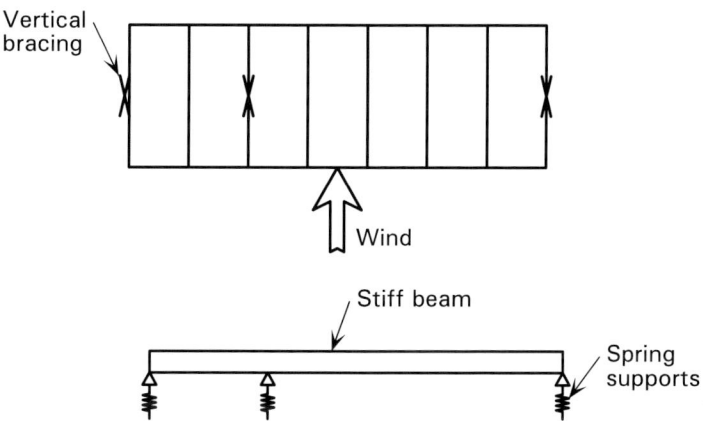

Figure 6.6 *Determination of bracing loads for asymmetric arrangement of bracing*

The spring stiffness of each bracing system should be calculated by applying horizontal loads to each bracing system and calculating the deflection. The spring stiffness (typically in mm/kN) can then be used to calculate the distribution of loads to each bracing system.

Vertical bracing is often constrained to be located within a cavity or in a particular vertical plane, for example around a lift shaft or stairwell. In all cases, eccentricity from the column axis should be minimised. Generally, the effect of eccentricity is not accounted for unless the line of the bracing is actually outside the plan area of the column. In this case, torques will be applied to the column which should be accounted for. Design of the columns in such circumstances is beyond the scope of this publication.

6.5 Vertical bracing design

Vertical bracing systems are generally modelled as cantilever trusses, as shown in Figure 6.7.

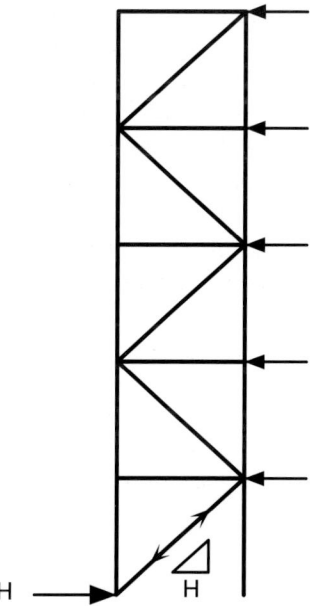

Figure 6.7 *Cantilever truss*

To size the diagonal members, it is often convenient to calculate the total lateral load on the structure, distribute this between the bracing systems and thus calculate the horizontal reaction at the foot of the bracing system. The resolved component of this horizontal shear can be used to size the bracing member, as shown in Figure 6.7

6.6 Bracing members and connections

Bracing members are often crossed flat steel members (designed in tension) or single hollow section (designed as struts). Other members may include angle sections (usually crossed and designed in tension only), channels or Universal Column sections, if heavily loaded. The local capacity of components within the connection can dominate the design of the member. Careful attention should be paid to how load is transferred to the bracing member, particularly if this is via thin elements such as webs.

Bracing connections generally involve a gusset plate to which the bracing member is bolted. If bracing is modelled assuming centreline intersection of all members, as shown in Figure 6.8, and this setting out is rigorously followed in the layout of members, it is not uncommon to find the actual bracing intersecting with either the beam or column some distance from the connection.

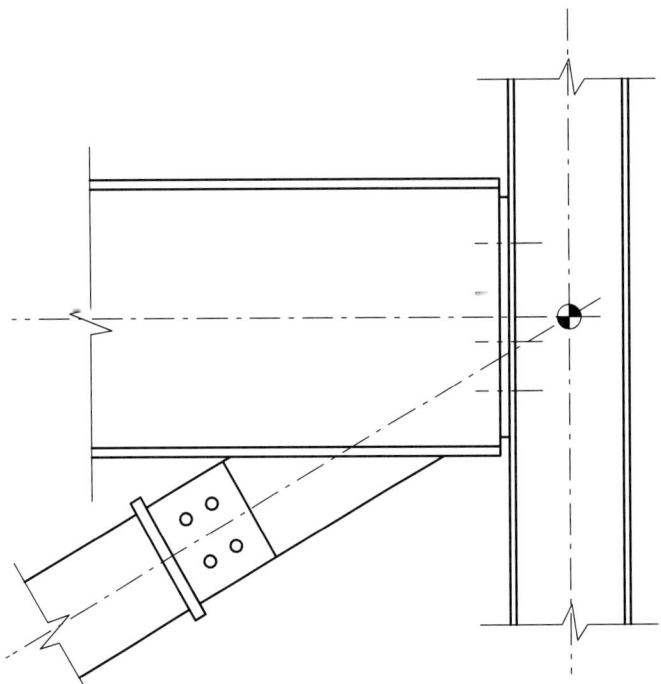

Figure 6.8 *Setting out of shallow bracing, with intersecting centrelines*

A better solution is to adjust the setting out point of the bracing, which usually results in a more compact gusset plate supported by both beam and end plate, as shown in Figure 6.9

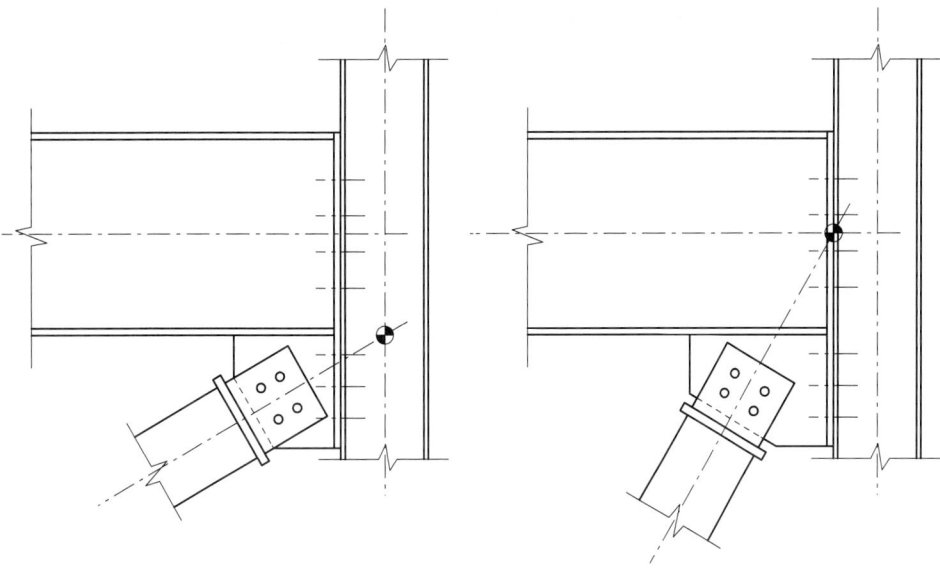

Figure 6.9 *Alternative setting out of bracing*

The setting out shown in Figure 6.9 does mean that additional moments are induced in the column, and this should be allowed for by re-analysing the column with rigid offsets to the bracing nodes, as shown in Figure 6.10.

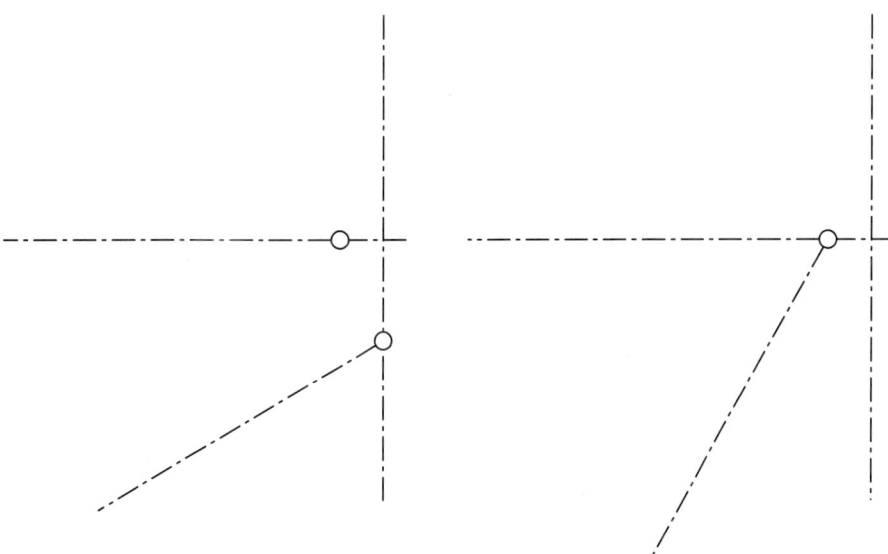

Figure 6.10 *Model for column analysis*

A careful consideration of load paths through the bracing systems is necessary to determine what proportion of load is transferred to or from the column at each node, and what loads remain within the bracing system. In the bracing systems shown in Figure 6.11, only a small force is transferred horizontally from the floor diaphragm through the column, as most of the horizontal component of the bracing force is internal to the bracing system. The transfer of the vertical component of the bracing force into the column is much more significant, and many more bolts would be expected for this purpose. Note that in the Pratt truss (Figure 6.11), the horizontal beam carries a significant axial force, and would need to be designed for such a force. The loads in the horizontal members of the Warren truss are significantly smaller.

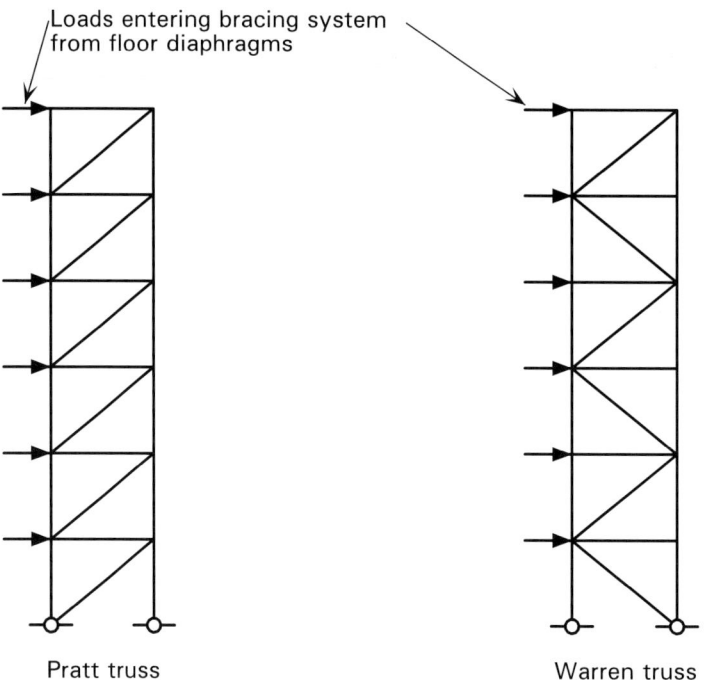

Pratt truss Warren truss

Figure 6.11 *Bracing systems*

Gusset plates such as that shown in Figure 6.12 are usually designed by considering the capacity of a section of the plate. For simplicity, the horizontal component of force is used to design the horizontal weld, and similarly the vertical component for the vertical weld. More complicated approaches, for example, considering the polar inertia of the weld group, when the line of the axial force is eccentric to the weld's centre of gravity, are possible, but beyond the scope of this guidance.

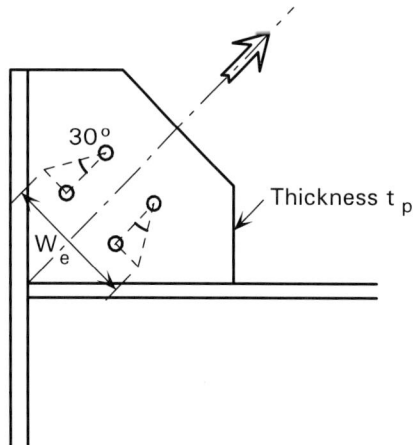

Figure 6.12 *Design of gusset plates in tension*

If gusset plates have to carry a compressive force, the plate should be of a generous thickness. Whilst no definite guidance exists, the thickness of the plate should be such that

$$t_p \geq \frac{w_e}{30}$$

where: t_p is the plate thickness and

 w_e is as shown in Figure 6.12.

The length of unstiffened plate should be minimised. If necessary, some consideration of the buckling capacity of the plate can be made, based on a pin-ended strut with length and breadth as shown in Figure 6.13.

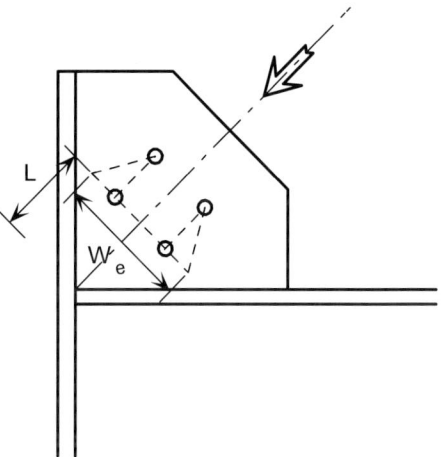

Figure 6.13 *Strut model for gusset plate in compression*

In almost all cases, the addition of gusset plates and the lengthening of the plates will invalidate the assumption that the connection is pinned. Experience has demonstrated however, that structures perform satisfactorily when still based on these simple solutions. If the setting out of the bracing varies significantly from centre-line intersection, columns should be modelled as shown in Figure 6.10.

6.6.1   Flat steel as bracing

The critical design check will be a tension check across the nett section, as shown in Figure 6.14.

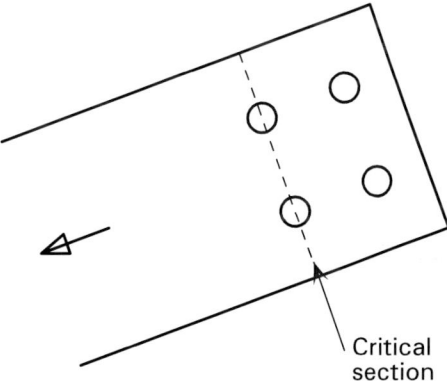

Figure 6.14 *Critical check for flat steel bracing*

Note that the net area depends on the hole size deducted (2 mm greater than bolt diameter) and that an effective net area can be calculated in accordance with Clause 3.4.3. Good practice is to choose a thickness that is greater than half the bolt diameter (if S275), so that bearing does not govern. The following table gives tensile capacity of typical flat bar bracing, in S275, assuming the bolts are in pairs across the bracing width.

Table 6.1 *Typical capacities of flat bracing*

Bolts	Flat	Capacity (kN)	Critical Element
4 M20	150 × 10	350	Plate
4 M20	150 × 12	368	Bolts
6 M20	200 × 12	552	Bolts
4 M24	180 × 15	528	Bolts
6 M24	200 × 15	733	Plate
6 M24	250 × 15	792	Bolts

Due to inevitable imperfections in the erection of a structure, one of the diagonal members in a pair of crossed flat steel braces is often not in tension at all, and tends to buckle laterally. Measures to ensure this does not happen include the deliberate fabrication of the bracing very slightly short, or the inclusion of some detail to adjust the length of the member (usually a turnbuckle or similar) once erected.

6.6.2 Hollow sections as bracing

Hollow sections will invariably be designed as struts, taking an effective length factor of 1.0, as typical gusset plate details are relatively flexible out of plane. The member length is generally calculated from the intersection of column and beam axis. Typical capacities in S275 are given below.

Table 6.2 *Typical capacities (as a strut) of circular hollow section bracing*

Length (m)	Size	Capacity (kN) (S275)
4.0	114 × 5	258
	139 × 5	413
	168 × 5	578
	193 × 8	1110
7.0	114 × 5	97.3
	139 × 5	177
	168 × 5	300
	193 × 8	670

Hollow sections are usually connected to a gusset plate via a tee, as shown in Figure 6.15. Larger loads are usually carried by a plate fitted into the section, as shown in Figure 6.16.

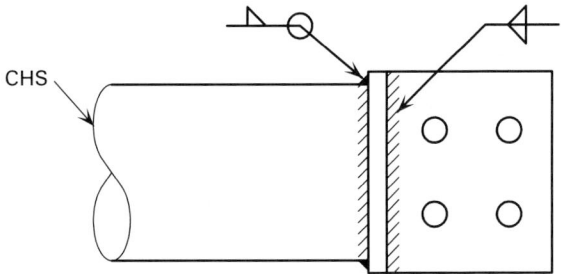

Figure 6.15 *Tee connection to hollow section*

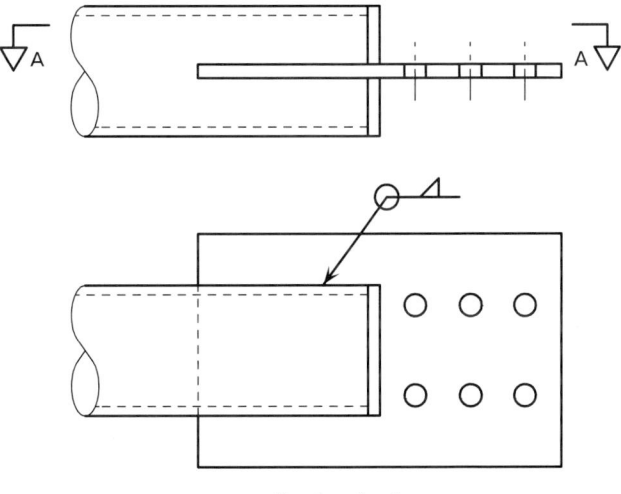

Section A - A

Figure 6.16 *Common detail for hollow sections carrying high loads*

For the tee type connection, the following checks are required:

- Bolts in shear and bearing.
- Tension check across the effective net area.
- Welds between the plated elements.
- Weld to the hollow section.

The weld checks for the detail in Figure 6.15 deserve special mention. The fillet weld between the plate elements is a transverse weld and this can be credited with an enhanced resistance in accordance with Clause 6.8.7.3 (the design strength of 220 N/mm^2 may be increased by 25%). Note that in calculating the length of weld, a length equal to the leg length should be deducted from each end of each weld.

When considering the weld between the tee and the member, only the weld falling in the 'stiff width' is considered, although the weld is continued at the same size all around the member. The stiff width (or effective breadth) may be calculated by assuming a spread at 1:25 through the plate, from the toe of the weld, as shown in Figure 6.17. The capacity of the member itself should also be checked on this basis – assuming only the part of the member falling within the effective breadth is carrying load.

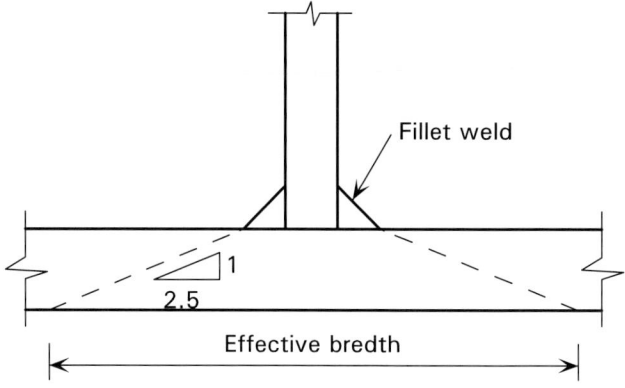

Figure 6.17 *Effective bredth*

If the plate is made thick enough to ensure the whole of the hollow section os effective, the weld between the plates (the tee) may also be considered as fully effective. If this is not the case, the weld should be designed on an effective length assuming a 1:2.5 distribution as shown in Figure 6.18

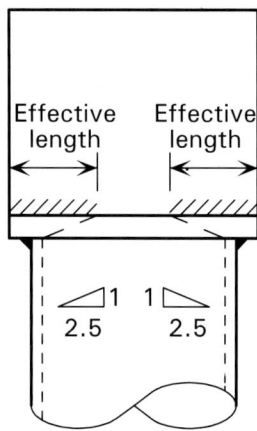

Figure 6.18 *Distribution of force into tee end connectors*

7 FRAME STABILITY

7.1 Introduction

This Section covers the explicit requirements of BS 5950-1 that all frames, including braced frames, be checked for sensitivity to second-order effects. The clarity of the frame stability checks is clouded by the use of Notional Horizontal Forces (NHF) in both the frame stability checks and as a minimum level of lateral load, and this leads to confusion. The distinction between the two quite separate uses of the NHF cannot be over-emphasised, and they should be treated as quite independent issues. The use of the NHF as a minimum level of lateral load is discussed in Section 7.3. The use of the NHF as part of the process to check frame stability is described in Section 7.5.

7.2 Frame behaviour

For the reason that we live in an imperfect world, columns that we happily assume to be vertical are never truly vertical. In every real column, a lateral component of the axial force is produced because the column is slightly inclined. The fact that the applied vertical loads are not directly above the foundation results in an eccentricity, which will lead to lateral movement and additional forces (and/or bending moments) that may not have been allowed for. Consider the braced frame shown in Figure 7.1.

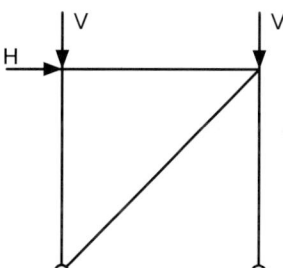

Figure 7.1 *Braced frame, undeformed.*

The tensile force in the bracing may be calculated by equating the horizontal component of the axial force to the applied horizontal load.

Due to the axial tension, the bracing extends, allowing the frame to move laterally, and producing an inclination in the columns, as shown in Figure 7.2. As the columns are now inclined, additional horizontal components of force must be resisted by the structure.

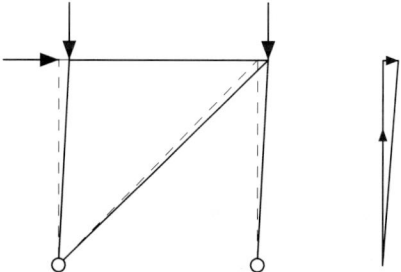

Figure 7.2 *Braced frame, deflected form*

Note that the horizontal components of the forces in the columns are proportional to the vertical loads, which demonstrates that frame stability is inextricably linked to vertical loads.

In many cases, the additional forces induced as the structure deflects are small enough to be ignored. In other cases, the forces are significant, and must be explicitly allowed for in design.

7.3 Minimum lateral load

To account for the imperfections that exist in the real world, BS 5950-1 requires in Clause 2.4.2.4 that a minimum level of lateral load be adopted, *if there are no other lateral loads*. This means that the minimum level of lateral load is *only* applied in load combination 1 (the gravity load combination). All other load combinations have lateral load, and no minimum is therefore applied.

The minimum level of lateral loading is taken as 0.5% of the factored vertical dead and imposed loads, and the loads are known as the Notional Horizontal Forces (NHF). In a braced frame, the NHF will be carried to the bracing system. The NHF are calculated by multiplying the whole floor area on each floor by the factored vertical load, multiplied by 0.5%. These loads are applied at each floor and roof level, and may be different at each level, particularly at roof level.

The Standard notes that the NHF do not contribute to the net reactions at the foundations. This is because no external loads are applied. However, NHF will contribute to the loads on *individual* foundations.

7.4 Minimum wind load

The Standard states in Clause 2.4.2.3 that a minimum wind load is taken as 1% of the factored dead load. Like the NHF, these loads are to be applied at floor level and roof level as point loads.

7.5 Frame stability checks

Sway sensitivity is measured by calculating the parameter λ_{cr}. A low value of λ_{cr} means the frame is more sensitive to second-order effects. BS 5950-1 allows λ_{cr} to be calculated by a simple process of applying Notional Horizontal Forces (NHF) to the otherwise unloaded frame, and calculating the lateral deflections. The NHF are 0.5% of the *factored* vertical loads, and are applied at floor and roof levels. Note that the NHF are load combination dependent.

λ_{cr} is calculated as $\dfrac{h}{200\delta}$

where: h is the storey height

δ is the difference in lateral deflection over the storey.

Each storey should be considered, with λ_{cr} for the whole structure taken as the lowest value found from any storey.

For braced frames, only the bracing system resists the lateral loads, so to calculate the lateral deflections the NHF need only be applied to the bracing system, as indicated in Figure 7.3. It is obviously essential that analysis of the bracing system uses the correct columns, beams and bracing members.

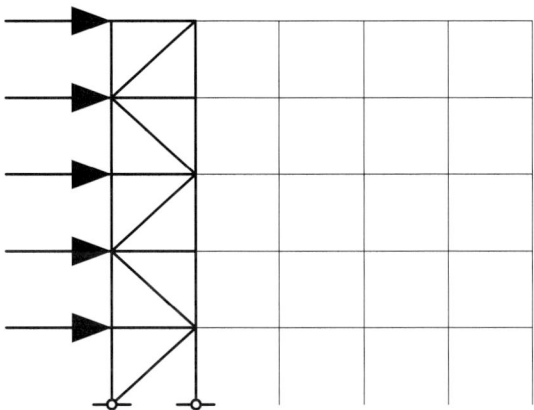

Figure 7.3 *NHF applied to the bracing system within a frame of simple construction*

The Standard describes two types of frames: the first type covers those frames where there are other stiffening features such as cladding but those features have been ignored when calculating λ_{cr}; the second type covers all other frames.

Frames with additional stiffness
In these cases, the analysis model used when calculating the lateral deflections is conservative. This is probably the most common case.

For this type of frame, the frame is classified as 'non-sway' if $\lambda_{cr} \geq 10$. This should strike a familiar chord with designers who checked frames to Section 5 of the 1990 Standard.

In the 1990 Standard, a clad continuous frame was classified as 'non-sway' if $\delta \leq \dfrac{h}{2000}$.

Some simple rearrangement reveals that passing the $h/2000$ check means $\lambda_{cr} \geq 10$.

All other frames – no additional stiffness or additional stiffness allowed for
In this type of frame, the analysis model is more realistic, i.e. the stiffening effects have been included, or the frame has no other stiffening, and the frame is always classified as 'sway-sensitive'.

7.6 Amplification to allow for sway-sensitivity

To allow for sway-sensitivity, BS 5950-1 requires the calculation of a simple amplifier, unless the frames are *very* sensitive, when the simple amplifier cannot be used.

The amplifier, k_{amp}, depends on the type of frame, as shown in Table 7.1 below:

Table 7.1 *Amplifier to allow for second-order effects*

Where additional stiffness has been ignored	All other frames
$k_{amp} = \dfrac{\lambda_{cr}}{1.15\lambda_{cr} - 1.5}$	$k_{amp} = \dfrac{\lambda_{cr}}{\lambda_{cr} - 1}$

If $\lambda_{cr} \leq 4$, the Standard requires that a second-order analysis is carried out, as the amplification approach cannot deal adequately with such sensitive frames.

Application of k_{amp}

Having calculated k_{amp}, the sway effects must be amplified. In a braced frame, the sway effects are those forces in the bracing system due to the lateral load. For simplicity, an easy approach is to multiply the lateral loads by k_{amp}, in addition to the partial factors for loads. Thus (1.4 × wind) will become (1.4 × k_{amp} × wind).

In load combination 1 (gravity loads) note that the only lateral loads are the NHF, and are usually small.

The application of k_{amp} will only affect the members that form the bracing system, usually the diagonal 'web' members and the columns forming the 'chords' of the vertical truss. k_{amp} will amplify all the force in any diagonal bracing members, but will only amplify part of the force in the columns, since only part of the force in the columns is due to the lateral loads – most is due to the vertical loads and is not amplified. Table 7.2 indicates how k_{amp} may conveniently be applied in each load combination.

Table 7.2 *Application of k_{amp} in different load combinations*

Load combination	Loading
1	1.4 × dead + 1.6 × imposed + k_{amp} × NHF
2	1.2 × dead + 1.2 × imposed + k_{amp} × 1.2 × wind
3	1.4 × dead + k_{amp} × 1.4 × wind
4	1.0 × dead + k_{amp} × 1.4 × wind

7.7 Frame stability and ULS load combinations

From Figure 7.2 it will be observed that the lateral component of force in the inclined column is proportional to the vertical load. Thus the NHF are calculated as 0.5% of the *factored* vertical load, and so vary with ULS load combination.

Since the NHF vary with load combination, so the calculation of λ_{cr} varies with load combination.

The maximum NHF are calculated in load combination 1 (Dead load plus imposed load), simply because the ultimate vertical loads are largest in this combination. In load combination 1, the frame will be most sensitive to second-order effects, indicated by the minimum value of λ_{cr}.

In other load combinations, the factored vertical loads are reduced. For example, when calculating the ultimate *vertical* loads, (1.4 × dead + 1.6 × imposed) becomes (1.2 × dead + 1.2 × imposed) in load combination 2. Thus the NHF are smaller in load combination 2 than load combination 1, which in turn means that the frame is more stable in load combination 2 than in load combination 1. A basic numerical demonstration of this effect is contained in Appendix 1.

The amplifier described in Section 7.6 is calculated from λ_{cr}, and therefore also varies with load combination.

The subtlety of this approach is that frames may well be classified as 'sway-sensitive' in load combination 1, but may be classified as 'non-sway' in other load combinations, meaning no amplification is necessary in those load combinations. A further subtle point is to note that although the largest (and most onerous) k_{amp} is determined from load combination 1, its application (see Table 7.2) is to a small lateral load – the NHF. In load combination 1 therefore, the effect of sway sensitivity is likely to be small, with no change to member sizes, as the bracing will have been sized on the wind loads.

A conservative approach is to check frame stability in load combination 1, and use the resulting amplifier in all load combinations, but economy may be improved by considering frame stability in each load combination separately.

An example of the frame stability calculations is presented in Reference [39] and further guidance is given in Reference [40].

8 ROBUSTNESS

8.1 Introduction

Structural codes include requirements for general robustness (or 'structural integrity') to ensure that frames have some minimum level of resistance to unexpected loading, and secondly to limit the extent of any collapse following an unexpected extreme event, such as an explosion. The requirements were introduced following the progressive partial collapse of a block of flats at Ronan Point, Newham, in 1968. Following a gas explosion in a lower flat, several floors above collapsed in a progressive manner, as each floor was unable to support the structure above, once its own support had been lost.

Robustness rules generally require the columns to be tied into the rest of the structure, so that the column cannot easily be removed, and that floor systems can develop catenary action following a partial collapse. Although the basic proposition of column removal might be as shown in Figure 8.1, it is clear that the robustness rules do not prescribe a design model for such circumstances – the reaction to the horizontal forces in Figure 8.1 is not addressed, for example. The robustness rules are best considered as prescriptive rules intended to produce structures that perform adequately in extreme circumstances, and are not meant to be fully described systems of structural mechanics. The illogical practice of designing certain connections for considerable load, yet not making provision to transfer the applied load any further, illustrates this point.

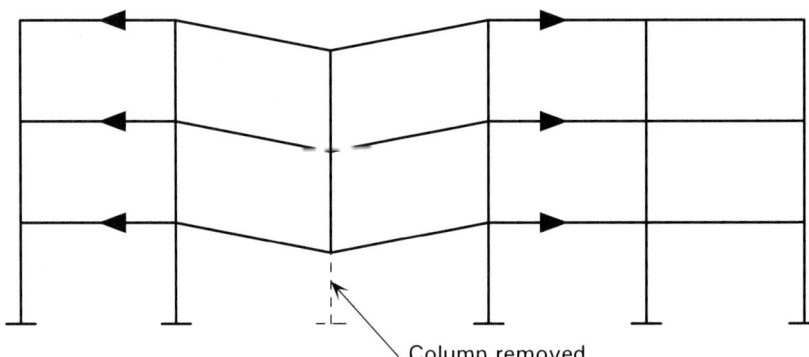

Figure 8.1 *Concept of robustness rules*

It should be noted that the requirements are not intended to ensure that the structure is still serviceable following some extreme event, but that damage is limited, and that progressive collapse is prevented. Thus BS 5950-1, Clause 2.4.5.1 notes that substantial permanent deformation of members and their connections is acceptable.

8.2 Structural integrity in BS 5950-1

In BS 5950-1, rules covering structural integrity are provided in Clause 2.4.5. Clause 2.4.5.2 specifies that columns must be tied in two directions, approximately at right angles, at each principal floor level. Clause 2.4.5.2 specifies that all ties (along the edges of the building and along each column line) should resist a factored load of 75 kN. In reality, this is achieved by any reasonable connection. All the standard details for simply supported beams[41] can carry at least 75 kN.

8.3 Avoiding disproportionate collapse

BS 5950-1 only specifies *how* buildings should be designed to avoid disproportionate collapse, and not *when* the provisions should be applied, prefacing the requirements with the phrase 'where Regulations stipulate'. The regulations referred to are the Building Regulations[2], which in the version published in 2000, prescribe that buildings above five storeys must be designed to avoid disproportionate collapse. It is expected that the Regulations will be revised in 2004, including changes to the scope of application.

In the draft revised Regulations, the cut-off below five storeys is removed, and robustness rules are applied depending on the category of structure, regardless of building height. If these rules remain when the revised Regulations are published, it is likely that the majority of structures will be subject to the more onerous requirements in Clause 2.4.5.3 of BS 5950-1. Broadly, this will mean that connections will need to be designed for a tying force equal to the shear reaction.

8.3.1 Designing for tying forces

Note that tying forces do not necessarily need to be carried by the steelwork frame. A composite concrete floor, for example, can be used to tie columns together, but must be designed to perform this function. Additional reinforcement may be required, and the columns (particularly edge columns) may need careful detailing to ensure the tying force is transferred between column and slab. Reinforcing bars around columns, or threaded bars bolted into the steel column itself have been successfully used.

If the tying forces are to be carried by the structural steelwork alone, note that the check for tying resistance is entirely separate to that for vertical load carrying capacity. The shear force and tying forces are never applied at the same time. Furthermore, the usual requirements that members and connections remain serviceable under applied loads are ignored when calculating resistance to tying, as 'substantial permanent deformation of members and their connections is acceptable'. Models to calculate the tying capacity of the industry standard connections are presented in the 'Green Book'[41], and are generally based on ultimate capacities, not yield stresses.

8.3.2 General tying

General tying to avoid disproportionate collapse is covered in Clause 2.4.5.3 (a) of BS 5950-1. Examination of the equations will show that the tying force is generally equal to the shear reaction, but not less than 75 kN.

By examining Figures 1 and 2 of BS 5950-1, it is clear that the tying requirements when disproportionate collapse is to be avoided (Figure 2 of BS 5950-1) are subtly more onerous than those described for all buildings in Clause 2.4.5.2 (Figure 1 of BS 5950). In any building, the columns must be tied, but secondary beams not connected to columns need not be designed as ties. This is shown in Figure 8.2 below. However, when designing to avoid disproportionate collapse, all beams must be designed as ties. This is shown in Figure 8.3.

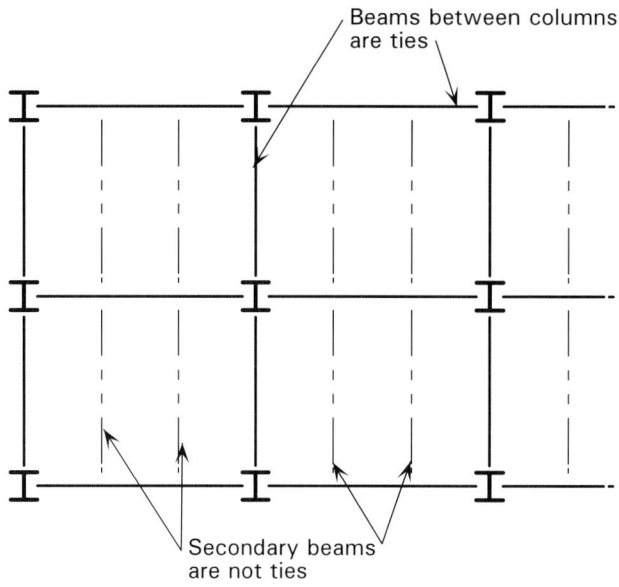

Figure 8.2 *Tying of all buildings (Clause 2.4.5.2)*

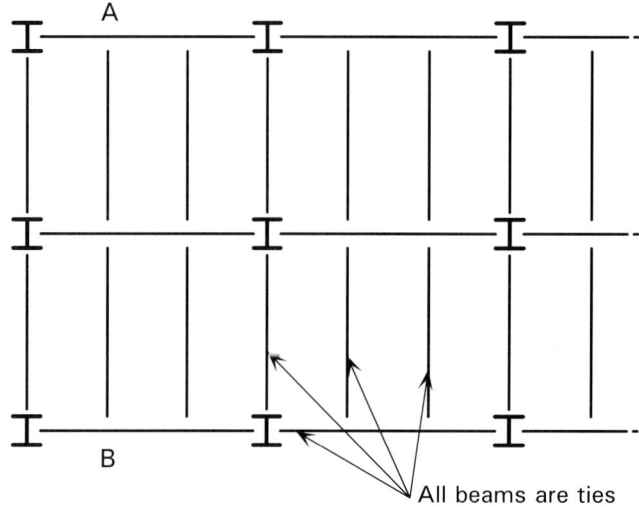

Figure 8.3 *Tying requirements when disproportionate collapse must be avoided*

Frequently, ties may be discontinuous, or have no 'anchor' at the end distant to the column. Two examples are shown in Figure 8.3, where at points A and B, there is no reaction to the tie force assumed in the beams running between points A and B. The connection is simply designed for the applied force. This situation is also common at external columns, where only the local design of the connection is considered. The column itself is not designed to resist the tying force.

8.3.3 Tying of edge columns

In BS 5950, Clause 2.4.5.3(b), states that the tying force is the larger of two forces; the tying force already described, or 1% of the factored axial load in the column at that point. It is straightforward to observe that usually, 1% of the factored load in the column only becomes the more critical load if there are a great many storeys; 100 storeys if all floors are identical. Columns carrying transfer trusses or similar massive loads may have high axial loads, and 1% of the factored axial load should always be considered.

In practice, members connected to columns will usually be acting as restraints to the column, and under the requirements of Clause 4.7.12 will have to be capable of resisting a force at least 1% of the axial force in the column. Members and connections conforming to Clause 4.7.12 are not allowed to suffer 'substantial permanent deformation' (see Section 8.3.1) and will therefore automatically satisfy the provisions of Clause 2.4.5.3(b).

8.3.4 Continuity of columns

Simply, the column splices must be capable of carrying an axial tension equal to the factored load imposed on the column by any single floor below the splice, down to the next splice or to the base.

In practice, this is not an onerous obligation, and most splices designed for adequate stiffness and robustness during erection are likely to be sufficient to carry the axial tying force. The 'Green Book'[41] has details of standard splices, and quotes axial tension capacities to simplify the design checks.

As an indication only, typical axial capacities for splices with cover plates (see Figure 5.4) are shown in Table 8.1 below:

Table 8.1 *Typical column splice tensile capacities (with flange cover plates)*

Upper Column	Lower Column	Tensile Capacity (kN)
203 UC	203 UC	736
254 UC	254 UC	736
305 UC	305 UC	1588
203 UC	254 UC	500

In all cases, the capacities quoted are limited by bolt shear, and adding additional bolts can easily increase capacities. In most cases, the shear capacity of the bolts is reduced in accordance with Clause 6.3.2.2, due to the presence of packing between the flange cover-plate and the upper column. Designers should refer to the 'Green Book' for detailed design checks.

8.3.5 Resistance to horizontal forces

If the structure must be designed to avoid disproportionate collapse, Clause 2.4.5.3(d) requires at least two sets of bracing in each orthogonal direction. No substantial part of the structure can be braced by only one set of bracing in the direction being considered. Thus, for buildings designed to avoid disproportionate collapse, the bracing arrangement in Figure 8.4 would not be satisfactory.

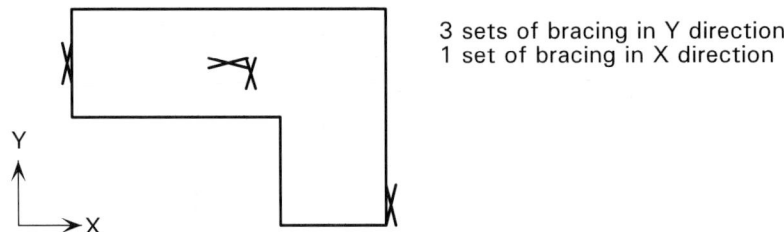

Figure 8.4 *Unsatisfactory bracing arrangement if disproportionate collapse is to be avoided*

8.3.6 Heavy floor units

Clause 2.4.5.3(e) requires that when concrete or other heavy floor units are used (as floors), they should be tied in the direction of their span. BS 5950-1 refers the designer to BS 8110-1. The intention of this clause is to prevent floor units or floor slabs simply falling through the steel frame, if the steelwork is moved or removed due to some major trauma. Slabs must be tied to each other over supports, and tied to edge beams. The tying forces may be calculated from BS 8110-1[42] Clause 3.12.3.4.

Tying across internal supports

If the precast units have a structural screed, it may be possible to use the reinforcement in the screed to carry the tie forces, as shown in Figure 8.5, or to provide additional reinforcing bars.

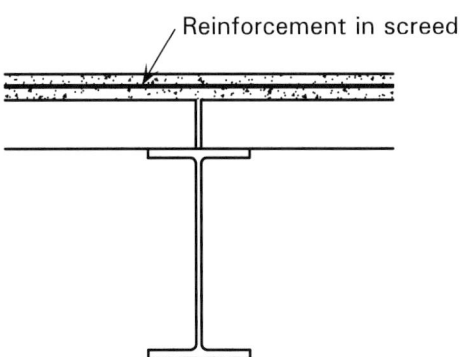

Figure 8.5 *Screed with reinforcement*

Alternatively, it may be possible to expose the voids in the precast planks and place reinforcing bars between the two units prior to concreting, as shown in Figure 8.6.

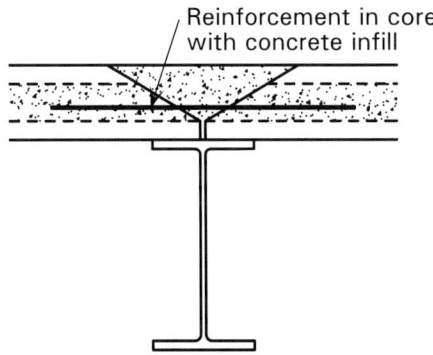

Figure 8.6 *Ties between hollow precast units*

Special measures will be needed where precast planks are placed on shelf angles as shown in Figure 8.7, and with *Slimflor* construction (see Section 4.4), unless the tie forces can be carried through the reinforcement in the screed, assuming this is above the top flange of the steelwork. When it is not possible to use reinforcement in the screed, straight reinforcement bars tying the precast units together is usually detailed to pass through holes drilled in the steel beam.

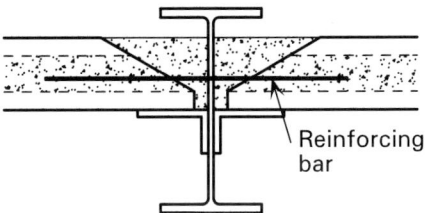

Figure 8.7 *Precast units on shelf angles*

Tying to edge beams

Anchorage is best accomplished by exposing the voids in the plank, and placing U-shaped bars around studs welded to the steelwork, as shown in Figure 8.8. In this Figure, the studs have been provided in order to achieve adequate anchorage; not for composite design of the edge beam. Other, more complicated solutions involve castellation of the plank edge, (often on site) so that the plank fits around the stud, and similar U-bars located in the voids prior to concreting.

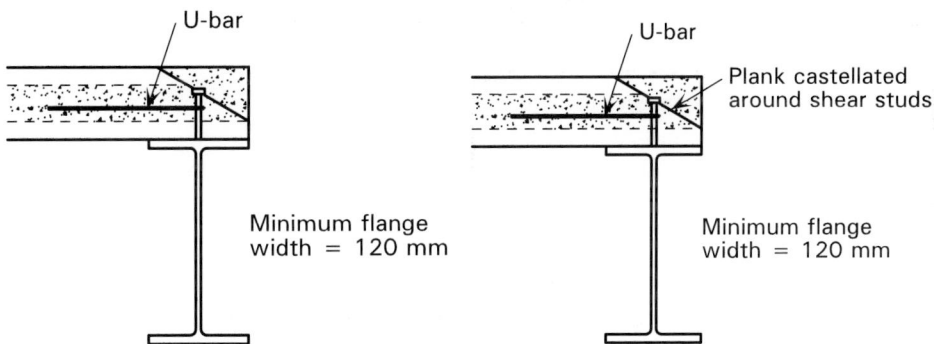

Figure 8.8 *Tying of precast planks to edge beams*

It should be noted that loading a beam on one side only produces significant torque in the beam itself, which may well be the critical design case. The eccentricity must be accounted for in design of the member, connections and columns.

In some circumstances, the floor units cantilever past the edge beam. Tying in these situations is not straightforward, and a solution must be developed in collaboration with the frame supplier and floor unit manufacturer.

9 FIRE RESISTANCE

9.1 Introduction

The fundamental concern for the structural designer when considering fire, is to design the structure such that its stability will be maintained for a reasonable period. This functional requirement is described in the Building Regulations. Ensuring the structural elements have a certain period of fire resistance (usually 30, 60 or 90 minutes) will generally be taken as evidence that the functional requirements for building stability in the event of a fire will be met. In England and Wales, appropriate periods of fire resistance are given in *Approved Document B*[43]. In Scotland, they are given in Part D of *Technical standards for compliance with the Building Standards*[44]. In Northern Ireland, it is *Technical Booklet E*[45] The appropriate periods of fire resistance for elements of structure depend on building type and height. The period of fire resistance is important, as the basic structure may be affected by the approach chosen to achieve the required resistance.

The fire resistance of an element of structure is judged against 3 criteria:

- Resistance to collapse, (R)
- Resistance to fire penetration, (E)
- Resistance to the transfer of excessive heat, (I)

Resistance to collapse is the ability of an element (beam, column, floor etc) to maintain loadbearing capacity or the ability not to collapse.

Resistance to fire penetration is the ability to maintain the integrity of the element against the penetration of flames and hot gases (this applies to fire-separating elements such as walls and floors)

Resistance to the transfer of excessive heat is the ability to provide insulation from high temperatures (which applies to fire separating elements such as walls and floors).

Steel beams and columns generally require applied fire protection. There are three main protection techniques:

- Cementitious spray, applied on site.
- Fire-resistant boards, fixed on site.
- Intumescent coatings, applied on- or off-site.

In some cases, instead of applying protective systems to the members, advantage may be gained by judicious choice of structural system, where the inherent fire resistance of the system means that for certain periods of fire resistance, no applied protection is required.

For floors constructed using a composite slab and composite beams, and requiring a fire resistance requirement of up to 60 minutes, a design approach is available which allows many floor beams to remain unprotected[19]. Many other types of steel beam and column also have up to 60 minutes fire resistance without applied protection[46], as illustrated in Figure 9.1. For fire resistance periods of up to 60 minutes, applied protection should therefore not be seen as the automatic solution.

Composite floors almost invariably do not require applied protection provided they are of sufficient depth to meet the insulation requirement (E) and have sufficient reinforcement to achieve the load bearing requirement (R)[47]. The required fire resistance period will generally be an important consideration when designing composite floors, as the area of mesh reinforcement required depends on the fire resistance required as well as the load carrying capacity.

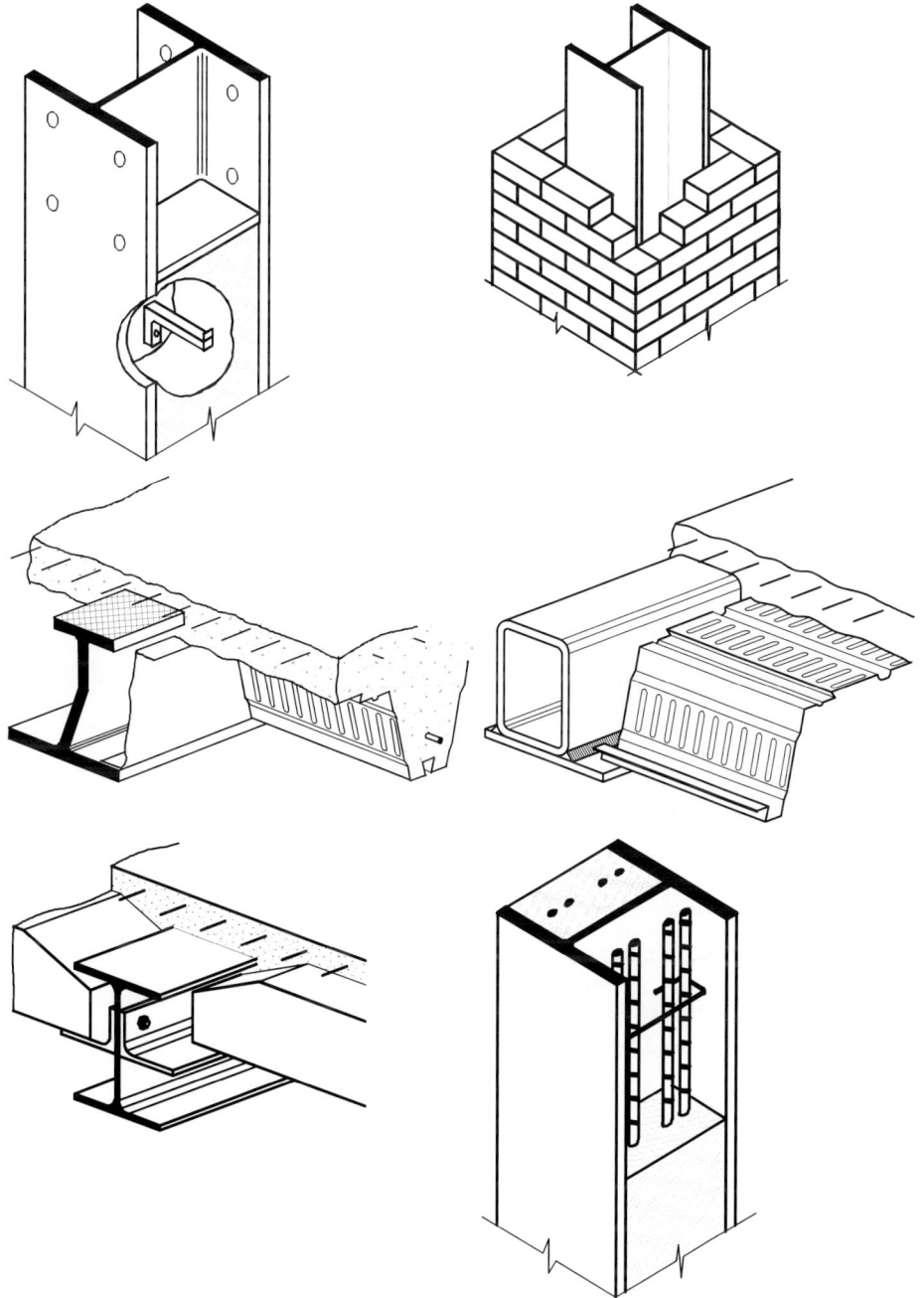

Figure 9.1 *Fire resistance without applied fire protection*

9.2 Periods of fire resistance

For construction in England and Wales, minimum periods of fire resistance for most elements of structure are set out in Appendix A, Table A2 of *Approved Document B*[43]. Scottish Building Regulations are performance based. However, minimum periods of fire are given by 'deemed to satisfy' provisions that can be found on Part D, Table 1[44]. For Northern Ireland, this information is given in *Technical Booklet E*[45] and has a similar format to Approved Document B. These minimum periods are summarised below for multi-storey buildings in Table 9.1. Definitions of height measurements are given in Figure 9.2.

Table 9.1 *Minimum periods (minutes) of fire resistance for elements of structure (based on Approved Document B, Table A2)*

Purpose of building	Basement storey including floor over		Ground or upper storey			
	Depth (m) of lowest basement		Height (m) of top floor above ground in building or separated part of building			
	≥ 10	< 10	≤ 5	≤ 18	≤ 30	> 30
Residential: flats and maisonettes	90	60	30	60	90	120
Office: Not sprinklered	90	60	30	60	90	NA
Office: Sprinklered	90	60	30	30	60	120

Note: Further categories of buildings are given in Approved Document B[43].

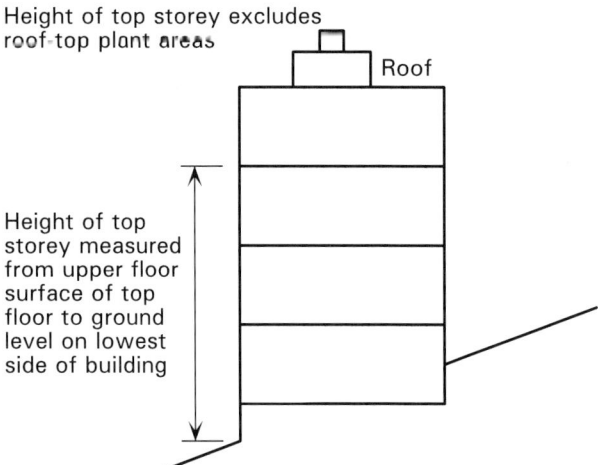

Figure 9.2 *Definitions of building height*

9.3 Fire protection systems

In the common situation where a period of fire resistance is prescribed, designers must consider how the period of fire resistance will be met, as the structural scheme is developed. If fire protection of the elements is necessary, the three common types of fire protection have important implications for cost, aesthetics and construction programme. A structural scheme cannot be developed in isolation.

Sprayed cementitious or gypsum based coatings

These relatively cheap forms of fire protection are usually applied on site, are excellent at coating complex fabrications, but have the disadvantages of a 'wet' trade on site. The application tends to be messy, and usually requires other trades to be excluded.

Boards and Blankets

Boards are made from a variety of materials, with different finishes. Columns and regular beams are often protected with boards. Blankets are often used around such items as trusses, where boards would be inefficient.

Intumescent coatings

Most intumescent coatings in common use are 'thin film', which swell and char on exposure to fire, protecting the steel. Intumescent coatings may be applied on site, but are often applied off-site, and consequently are off the critical site construction programme. Complex fabrications such as cellular beams or plate girders are usually protected with intumescent coating.

9.4 Sources of further advice

The following resources are recommended further reading:

- *Structural fire safety: A handbook for architects and engineers*[9]
  This publication presents a useful overview of the subject, without detailed design guidance.

- *Design of steel-framed buildings without applied fire protection*[46]
  This publication describes the ways in which up to 60 minutes fire resistance can be achieved without protection. The publication also contains worked example calculations of the fire engineering design of:
    - A *Slimflor* beam
    - An ASB
    - A beam with shelf angles
    - A partially encased beam
    - A partially encased column
    - A concrete filled structural hollow section
    - A shallow composite slab
    - A deep composite slab

- *Fire safe design – A new approach to multi-storey steel-framed buildings*[19].
  This publication describes the design approach for structures with composite floors and a required fire resistance period of up to 60 minutes. By following the guidance, many floor beams can be left unprotected, offering considerable advantage.

10 REFERENCES

1. BS 5950: Structural use of steelwork in building
   BS 5950-1:2000 Code of practice for design: rolled and welded sections
   British Standards Institution, 2001

2. Building Regulations 1991 - Approved Document A – Structure
   A3 and A4 Disproportionate collapse (incl Amds 1992; 1994)
   The Stationery Office, 1991

3. BRITISH COUNCIL FOR OFFICES
   BCO Guide 2000 Best practice in the specification for offices
   BCO, 2000
   http://www.bco-officefocus.com

4. CIMSTEEL
   Design for construction (P178)
   The Steel Construction Institute, 1997

5. McKENNA, P. D. and LAWSON, R. M.
   Interfaces: Design of steel-framed buildings for service integration (P166)
   The Steel Construction Institute, 1997

6. MITCHELL, S. HEYWOOD, M. and HAWKINS, G.
   Service co-ordination with structural beams. Guidance for a defect-free interface (IPE2)
   The Steel Construction Institute and BSRIA, 2004

7. HICKS, S. J. and DEVINE, P. J.
   Design guide on the vibration of floors. Second Edition (P076 2nd Ed)
   The Steel Construction Institute, 2004 (*to be published*)

8. HICKS, S. J. and DEVINE, P. J.
   Design guide on the vibration of floors in hospitals
   The Steel Construction Institute, 2004

9. HAM, S. J., NEWMAN, G. M., SMITH, C. I. and NEWMAN, L. C.
   Structural fire safety: A handbook for architects and engineers (P197)
   The Steel Construction Institute, 1999

10. Building Regulations 2000 - Approved Document E
    Resistance to the passage of sound
    E1: Protection against sound from other parts of the building and adjoining buildings
    E2: Protection against sound within a dwelling-house etc.
    The Stationery Office, 2003

11. BS 8233:1999 Sound insulation and noise reduction for buildings – Code of Practice
    British Standards Institution, 1999

12 GORGOLEWSKI, M. T. and LAWSON, R. M.
    Acoustic performance of *Slimdek* (P321)
    The Steel Construction Institute 2003

13 GORGOLEWSKI, M. T. and LAWSON, R. M.
    Acoustic performance of composite floors (P322)
    The Steel Construction Institute, 2003

14 GORGOLEWSKI, M. T. and LAWSON, R. M.
    Acoustic performance of light steel-framed systems (P320)
    The Steel Construction Institute, 2003

15 Building Regulations 2000 - Approved Document L2
    Conservation of fuel and power - L2 Conservation of fuel and
    power in buildings other than dwellings (2002 edition)
    incl AMD 1 - May 2002
    The Stationery Office, 2002

16 BS 6399-1:1996 Loading for buildings. Code of practice for dead and
    imposed loads
    British Standards Institution, 1996

17 BS 6399-3:1988 Loading for buildings. Code of practice for imposed roof
    loads
    British Standards Institution, 1988

18 BROWN, D. G.
    Recommended application of BS 6399-2 (ED001)
    The Steel Construction Institute, 2001
    (Available online to SCI members; www.steelbiz.org)

19 NEWMAN, G. M., ROBINSON, J. T. and BAILEY, C. G.
    Fire safe design: A new approach to multi-storey steel-framed buildings
    (P288)
    The Steel Construction Institute, 2000

20 COUCHMAN, G. H., MULLETT, D. L. and RACKHAM, J. W.
    Composite slabs and beams using steel decking: Best practice for design
    and construction (P300); *currently under revision*
    The Steel Construction Institute, 2000

21 LAWSON, R. M.
    Design of composite slabs and beams with steel decking (P055)
    The Steel Construction Institute, 1989

22 ASSOCIATION FOR SPECIALIST FIRE PROTECTION and THE
    STEEL CONSTRUCTION INSTITUTE
    Fire protection for structural steel in buildings. Third edition
    ASFP/ SCI/ FTSG, 2002

23 CORUS
Slimdek manual
Corus, 2001

24 LAWSON, R. M., MULLETT, D. L. and RACKHAM, J. W.
Design of asymmetric *Slimflor* beams using deep composite decking (P175)
The Steel Construction Institute, 1997

25 MULLETT, D. L.
Design of RHS *Slimflor* edge beams (P169)
The Steel Construction Institute, 1997

26 AD 269: The use of intumescent coatings for the fire protection of beams with circular web openings
Advisory Desk in New Steel Construction, vol 11(6), 2003

27 SIMMS, W. I.
RT983: Interim guidance on the use of intumescent coatings for the fire protection of beams
The Steel Construction Institute, 2004

28 WARD, J. K.
Design of composite and non-composite cellular beams (P100)
The Steel Construction Institute, 1994

29 MULLETT, D. M.
Slim floor design and construction (p110)
The Steel Construction Institute, 1992

30 HICKS, S. J. and LAWSON, R. M.
Design of composite beams using precast concrete slabs (P287)
The Steel Construction Institute, 2003

31 Steelwork design guide to BS 5950-1:2000
Volume 1: Section properties and member capacities. 6th Edition (P202)
The Steel Construction Institute and The British Constructional Steelwork Association, 2001

32 WAY, A. G. J., and SALTER, P. R.
Introduction to steelwork design to BS 5950-1:2000 (P325)
The Steel Construction Institute, 2003

33 AD 243: Splices within unrestrained lengths
Advisory Desk in New Steel Construction, vol 8(6), 2000

34 National structural steelwork specification for building construction (4th edition)
The British Constructional Steelwork Association and The Steel Construction Institute, 2003

35 Joints in steel construction: Composite connections (P213)
The Steel Construction Institute, 1998

36 Joints in steel construction: Moment connections (P207)
The Steel Construction Institute and The British Constructional Steelwork Association, 1995

37 Holding down systems for steel stanchions
Constrado, BCSA and Concrete Society, 1980

38 DAVISON, B. and OWENS, G. W. (*Editors*)
Steel designers' manual (6th edition)
Blackwell Publishing, 2003

39 HEYWOOD, M. D. and LIM, J. B.
Steelwork design guide to BS 5950-1:2000
Volume 2: Worked examples (P326)
The Steel Construction Institute, 2003

40 BROWN, D. G.
Multi-storey frame design
In New Steel Construction, vol 10(6), 2002

41 MALIK, A. S. (*Editor*)
Joints in steel construction: Simple connections (P212)
The Steel Construction Institute and The British Constructional Steelwork Association, 2002

42 BS 8110 Structural use of concrete.
BS 8110-1:1997 Code of practice for design and construction
British Standards Institution, 1997

43 Building Regulations 1991 - Approved Document B – Fire Safety - (includes 1992 amendments - now superseded by 2000 edition)
See also: Amendments 2002 to Approved Document B (Published by ODPM)
The Stationery Office, 1991

44 The Building Standards Amendment (Scotland) Regulations 2001
Technical Document D, Structural Fire Precautions
Scottish Executive, 2002

45 The Building Regulations (Northern Ireland)
Technical Booklet E Fire Safety (as amended 2000)
The Stationery Office, Belfast

46 BAILEY, C. G., NEWMAN, G. M. and SIMMS, W. I.
Design of steel-framed buildings without applied fire protection (P186)
The Steel Construction Institute, 1999

47 NEWMAN, G. M.
The fire resistance of composite floors with steel decking. Second edition (P056)
The Steel Construction Institute, 1991

APPENDIX A   Demonstration of frame stability calculations

A.1   Frame Stability Example

The principles in Section 7.5 can be demonstrated by considering the simple frame in Figure A.1. The stability of this frame will be examined under increasing levels of axial load. The horizontal load will not be changed. Note that all loads are quoted at the ultimate limit state (ULS).

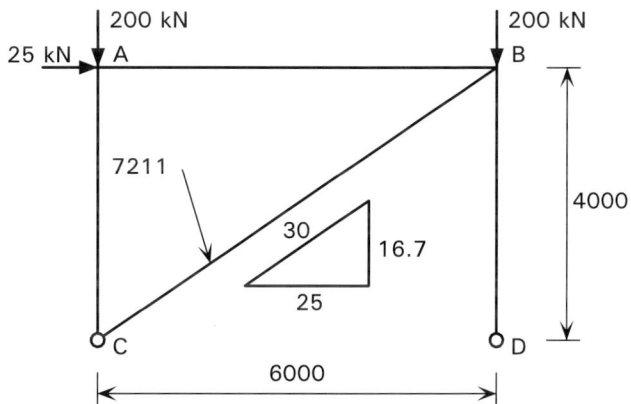

Figure A.1 *Simple braced frame*

For each level of axial load, an indication of the actual increases in force will be calculated, and compared with the rules in the Standard. It is assumed that the frame is clad, though this additional stiffening will not be allowed for.

A.1.1   Axial force of 200 kN in columns

The force in the diagonal bracing can be calculated as 30 kN.

Assuming a bracing area of 200 mm²,

Stress in bracing, $\sigma = \dfrac{30 \times 10^3}{200} = 150 \text{ N/mm}^2$

The strain in the bracing $\varepsilon = \dfrac{\sigma}{E} = \dfrac{150}{205 \times 10^3} = 7.3 \times 10^{-4}$

Since the original length of the bracing was 7211 mm, the extension is $7211 \times 7.3 \times 10^{-4} = 5.3$ mm.

The new length of the bracing is therefore $7211 + 5.3 = 7216.3$ mm.

The exaggerated shape of the deformed structure is shown in Figure A.2.

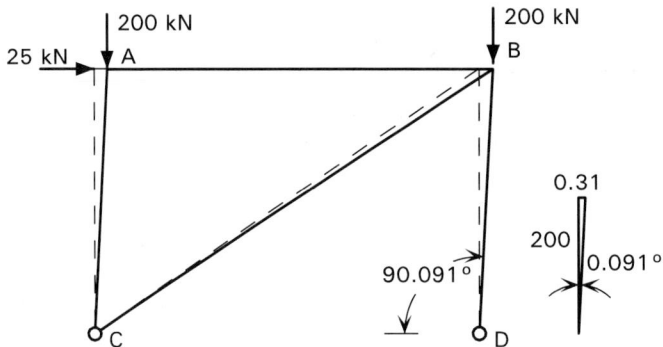

Figure A.2 *Deformed structure*

Using the cosine rule, the angle CDB can be calculated as 90.091°, indicating that the columns are now inclined from the vertical by 0.091°. It is assumed that the column does not shorten under axial load.

With an axial load of 200 kN in the columns, the lateral component is 200 tan 0.091 = 0.31 kN per column or 0.6 kN total.

This additional component of horizontal force is $\frac{0.6}{25}$ or 2.5%, and small enough to be ignored.

λ_{cr}, with 200 kN in the columns

The notional horizontal forces are 0.5% of the factored vertical loads, and in this case are therefore $\frac{0.5}{100} \times 2 \times 200 = 2$ kN.

Following the process demonstrated above, it is found that the frame sways by 0.5 mm under the action of the NHF.

λ_{cr} can then be calculated, being $\frac{h}{200} = \frac{4000}{200 \times 0.5} = 40$

Clearly this exceeds 10 by a considerable margin, and the frame would be classed as non-sway.

A.1.2   Axial force of 750 kN in each column

Since the horizontal load has not changed, the deformed shape of the structure is as shown in Figure A.2. Because the axial load in the columns has increased, the horizontal component per column is now 1.2 kN, or 2.4 kN total.

The additional component of horizontal force is $\frac{2.4}{25}$ or 10%.

λ_{cr}, with 750 kN in the columns

The NHF increase to $\frac{0.5}{100} \times 2 \times 750 = 7.5$ kN.

The resulting sway is 1.9 mm.

$$\lambda_{cr} = \frac{h}{200\delta} = \frac{4000}{200 \times 1.9} = 10.5$$

Since $\lambda_{cr} > 10$ the frame is (just) non-sway. Note that the frame has cladding, which would serve to reduce the calculated deflection.

A.1.3   Axial force of 1200 kN in each column

Under this loading, the additional horizontal component in each column is 1.9 kN, making a total addition of $\frac{2 \times 1.9}{25}$ or 15%.

λ_{cr}, with 1200 kN in the columns

The NHF increase to $\frac{0.5}{100} \times 2 \times 1200 = 12$ kN.

The resulting sway is 3 mm.

$$\lambda_{cr} = \frac{h}{200\delta} = \frac{4000}{200 \times 3} = 6.67$$

As this is less than 10, the frame is sway-sensitive.

Since the stiffening effect of cladding has been ignored, the amplifier k_{amp}, is given by:

$$k_{amp} = \frac{\lambda_{cr}}{1.15\lambda_{cr} - 1.5} = \frac{6.67}{1.15 \times 6.67 - 1.5} = 1.08$$

Whilst this amplification is less than the 15% increase calculated above, note that cladding would have reduced the calculated deflections.

If no stiffening were present, the amplifier k_{amp}, would be given by:

$$k_{amp} = \frac{\lambda_{cr}}{\lambda_{cr} - 1} = \frac{6.67}{6.67 - 1} = 1.18, \text{ or an 18\% increase.}$$

A.1.4   Calculation summary

For convenience, the calculations of the preceding sections are summarised in Table A.1 below:

Table A.1 *Summary of stability calculations (cladding ignored in sway calculation, $k_{amp} = \dfrac{\lambda_{cr}}{1.15\lambda_{cr} - 1.5}$)*

200 kN Each Column		750 kN Each Column		1200 kN Each Column	
Calculated Increase 2.5%	NHF 2 kN	Calculated Increase 10%	NHF 7.5 kN	Calculated Increase 15%	NHF 12 kN
	$\lambda_{cr} = 40$		$\lambda_{cr} = 10.5$		$\lambda_{cr} = 6.67$
	Non-sway		Non-sway		Sway-sensitive
	No Amplification		No Amplification		$k_{amp} = 1.08$

If the frame had no additional stiffening, Table A.2 shows the modified Figures.

Table A.2 *Summary of stability calculations (bare frames and frames where cladding is taken into account, $k_{amp} = \dfrac{\lambda_{cr}}{\lambda_{cr} - 1}$)*

200 kN Each Column		750 kN Each Column		1200 kN Each Column	
Calculated Increase 2.5%	NHF 2 kN	Calculated Increase 10%	NHF 7.5 kN	Calculated Increase 15%	NHF 12 kN
	$\lambda_{cr} = 40$		$\lambda_{cr} = 10.5$		$\lambda_{cr} = 6.67$
	Sway-sensitive		Sway-sensitive		Sway-sensitive
	$k_{amp} = 1.025$		$k_{amp} = 1.10$		$k_{amp} = 1.18$

Note that these simplistic calculations have ignored other effects, such as the axial shortening of the column or the beam.